T0091651

Algebra
for Parents
A Book for Grown-Ups About
Middle School Mathematics

Other World Scientific Titles by the Author

Arithmetic for Parents: A Book for Grown-Ups About Children's Mathematics
Revised Edition
ISBN: 978-981-4602-89-1
ISBN: 978-981-4602-90-7 (pbk)

Mathematics, Poetry and Beauty
ISBN: 978-981-4602-93-8
ISBN: 978-981-4602-94-5 (pbk)

Circularity: A Common Secret to Paradoxes, Scientific Revolutions
and Humor
ISBN: 978-981-4723-67-1
ISBN: 978-981-4723-68-8 (pbk)

Algebra
for Parents
A Book for Grown-Ups About
Middle School Mathematics

Ron Aharoni
Technion, Israel institute of Technology, Israel

World Scientific

NEW JERSEY · LONDON · SINGAPORE · BEIJING · SHANGHAI · HONG KONG · TAIPEI · CHENNAI · TOKYO

Published by

World Scientific Publishing Co. Pte. Ltd.

5 Toh Tuck Link, Singapore 596224

USA office: 27 Warren Street, Suite 401-402, Hackensack, NJ 07601

UK office: 57 Shelton Street, Covent Garden, London WC2H 9HE

Library of Congress Cataloging-in-Publication Data
Names: Aharoni, Ron, author.
Title: Algebra for parents : a book for grown-ups about middle school mathematics /
    Ron Aharoni, Technion, Israel Institute of Technology, Israel.
Description: New Jersey : World Scientific, [2021]
Identifiers: LCCN 2020041069 | ISBN 9789811209222 (hardcover) | ISBN 9789811210747 (paperback) |
    ISBN 9789811209239 (ebook for institutions) | ISBN 9789811209246 (ebook for individuals)
Subjects: LCSH: Algebra--Study and teaching. | Algebra--Study and teaching (Middle school)
Classification: LCC QA152.3 .A24 2021 | DDC 512.9071/2--dc23
LC record available at https://lccn.loc.gov/2020041069

British Library Cataloguing-in-Publication Data
A catalogue record for this book is available from the British Library.

For any available supplementary material, please visit
https://www.worldscientific.com/worldscibooks/10.1142/11524#t=suppl

Desk Editor: Liu Yumeng

Typeset by Stallion Press
Email: enquiries@stallionpress.com

Contents

Introduction

Mathematics is for the lazy. It means letting the principles do the work for you.

George Polya, a Hungarian-American mathematician and educator.

For the lazy? You must be kidding, Professor Polya. Everybody knows that mathematics is the hardest academic subject! Go entice others with promises of easy life.

But Polya means precisely what he says. In fact, this is true for all abstract thinking. A short-term effort results in a long term economy. We invent an abstraction, say "cat", and using it we know what to expect from every cat we meet, instead of learning anew how to behave with such a creature every time we meet one. A one-time investment bears long range fruit.

So, why does Polya speak of mathematics, among all abstractions? Of course — because he is a mathematician. But not only. Mathematics is the ultimate abstraction. It abstracts the most basic thought processes. First and foremost — numbers. The number "3" is an abstraction from three fingers, three apples and three cats. We realize that when it comes to arithmetic operations, it doesn't matter what we are counting. If 2 apples and 3 apples are 5 apples, the same will be true for pencils and books. We summarize it concisely: $2 + 3 = 5$. It is true for any type of objects, and at any time. Economy at its best.

Learning the abstraction "3" is like learning to recognize a certain person — a useful feat in itself. A baby learns to recognize that the figure that appears on and off is actually the same person. The next step of abstraction is making generalizations on all human beings. All people have heads and legs. In mathematics, this corresponds to stating generalizations about all numbers. This is algebra. Algebra says things like "adding 1 to any number enlarges it". This is a much more demanding abstraction. No wonder that we wait with it for adolescence — the age in which the power of abstraction makes a leap forward. The famous x's are the tool for that. See, for example, how compactly a formula says the same thing:

$$x + 1 > x.$$

Note how short is the formula in comparison with the verbal formulation.

This book is a step by step guide to the teaching of algebra. Alongside the mathematical material, it tells of some teaching principles. Real understanding must be active, and hence I included some exercises. I chose to put only basic ones — more intricate exercises can be found in textbooks.

Like all abstractions, algebra is beautiful. For those who are looking for the beauty side in mathematics, I inserted chapters called "a bit beyond" at places. These are topics that are not usually studied at school, and are there mainly for their beauty.

Two teaching principles

Two teaching principles are so prominent, that I want to mention them already now. One is inventing problems on your own: invent an equation having 2 as a solution; find a quadratic equation whose solutions are 0 and 1; invent a problem for which the solution is "the travel time was 2 hours". Inventing problems gives better mastery of the subject than solving them. The other principle relates to thought in general, not only in teaching: look for simple examples. What is the simplest quadratic equation? What is the simplest travel problem you can formulate? A simple example strips the topic of its less relevant details, and leaves its essence.

Part 1

Variables and Unknowns

One merit of poetry few people will deny: it says more and in fewer words than prose.

Voltaire, Dictionnaire philosophique portatif

("A Philosophical Dictionary") (1764), Poets.

"Poetry" in German is "Dichtung", which means "compression" — indeed, one of its best known properties. In this, mathematics is a worthy competitor. Remember? "Mathematics is for the lazy". And a main agent in its economy is the formulas, that say things concisely and precisely. In other words, algebra.

To introduce this idea to students, I sometimes tease them. In the 11-th or 12-th grade I ask them "What is algebra?". They have been studying algebra for at least four years by then, and still they stammer. The best I can draw from them is "it is x's".

I write a few pairs of equalities:

$$10 \times 10 = 100 \quad 9 \times 11 = 99$$
$$8 \times 8 = 64 \quad\quad 7 \times 9 = 63$$
$$5 \times 5 = 25 \quad\quad 4 \times 6 = 24$$

Can you find regularity here? — I ask them. To check their understanding, I ask them to add one more example of their own.

> A basic teaching principle: examples. Abstractions are understood only through examples. And best are examples invented by the students themselves.

What is the simplest example? There are two contestants for this title:

$$1 \times 1 = 1, \quad 0 \times 2 = 0 \quad \text{and} \quad 0 \times 0 = 0, \ (-1) \times 1 = -1$$

> One should first look for the simplest example possible. There is no such thing as "too simple".

Here is an attempt of a verbal formulation: the first member of each pair is the product of a number by itself. The second is the product of the number plus 1 by the number minus 1. And the result is 1 less than in the first pair.

Did you follow? Even I can hardly follow, though I just wrote it. Here is a shorter version: "a number times itself is the number plus 1, times the number minus 1, plus 1". Better, but still cumbersome. Why is it that thought stumbles so badly in its way to the light of explicit formulation?

The reason is that we need a name for the number. Without names, it is like referring as "that person" to the hero of a story. And best is a short name, like a letter. Say, a. The law is then:

$$a \times a = (a-1) \times (a+1) + 1$$

Now one can understand.

Identities

What does "a" signify in the last formula? Of course, any number. The rule is valid, as we deduced by experimentation, for all numbers.

A letter-name denoting a general number is called a "variable". It "varies" over all possible numbers. This is a first use of algebra: formulating an equality that is true for any number. Such an equality is called an "identity". Algebra uses letters to formulate identities. Of course, it does not matter what letter is used. We could just have written $b \times b = (b-1) \times (b+1) + 1$ as well.

Use the identity $a \times a = (a-1) \times (a+1) + 1$ to calculate 19×21 and 99×101.

More Identities

Here are a few more identities, simpler and better known:

$$x + 0 = x$$
$$0 \cdot x = 0$$
$$x - x = 0$$
$$a + b = b + a$$
$$\frac{b}{b} = 1$$

Complete the following identities: $a + 0 =$; $a + b - b =$; $\frac{1}{1/x} =$;
$a \times 1 =$.

A tip for guessing the identity: substitute numbers for the variables. For example, putting $x = 4$ in $\frac{1}{1/x}$ gives $\frac{1}{1/4}$, which is 4 (an explanation why is given in one of the next chapters). It is easy to guess from this that $\frac{1}{1/x} = x$. By the way, here x is not totally general — it cannot be 0, since division by 0 is not permitted.

Another example: look at the following equalities:

$$\frac{1}{2} - \frac{1}{3} = \frac{1}{6}, \quad \frac{1}{3} - \frac{1}{4} = \frac{1}{12}, \quad \frac{1}{4} - \frac{1}{5} = \frac{1}{20}, \quad \frac{1}{5} - \frac{1}{6} = \frac{1}{30}$$

If you do not remember how to subtract fractions, just take my word that I calculated right. Look at the equality $\frac{1}{5} - \frac{1}{6} = \frac{1}{30}$. In the denominator of the right hand side there is the product of the two left hand side denominators. Calling the first denominator (in this case 5) m, the second denominator is $m + 1$, the rule is then:

$$\frac{1}{m} - \frac{1}{m+1} = \frac{1}{m \times (m+1)}$$

Check that the equality $\frac{1}{4} - \frac{1}{5} = \frac{1}{20}$ fits this formula. What is m here? Generalize by a formula the rule exemplified by the following:

$$10 \times 10 = 100 \quad 8 \times 12 = 96$$
$$8 \times 8 = 64 \quad 6 \times 10 = 60$$
$$5 \times 5 = 25 \quad 3 \times 7 = 21$$

Solution: $(x - 2) \times (x + 2) = x^2 - 4$. Similar to the law in the beginning of this chapter.

Put a number for the question mark, so as to make the following equality an identity: $(x - 3) \times (x + 3) = x^2 - ?$
Generalize the following equalities, and formulate a corresponding identity:

$$3^2 + 4^2 + 12^2 = 13^2$$
$$5^2 + 6^2 + 30^2 = 31^2$$
$$10^2 + 11^2 + 110^2 = 111^2$$

Solution: $n^2 + (n+1)^2 + (n \times (n+1))^2 = (n \times (n+1) + 1)^2$.

Unknowns

When the police are searching for the identity of a criminal, they give him a nickname, like "the motorbike robber". This is also done in algebra. Sometimes we are looking for a number that is not known, but we have some information on it. We give it a name, which as usual is a letter, and this enables referring to it conveniently. In this role, the letter is called an "unknown".

Albert Einstein related how he was infected with love for mathematics as a child after reading a famous riddle, inscribed on the grave of the great ancient Greek mathematician Diophantus. Adhering to the basic teaching rule of starting from the simple, I will replace this riddle by a simpler problem of its type, leaving the Diophantus problem for later.

> David's age is 3 times that of his brother John, and in 5 years his age will be 2 times that of John. How old are David and John?

Let us call John's age x (we could also give a letter-name, say y, to the age of David, but this would result in dealing with fractions, since then John's age is $\frac{1}{3}y$). Then David's age is $3x$. In 5 years time David's age will be $3x + 5$ and John's age will be $x + 5$. The fact that David is going to be twice as old as John can be written as:

$$3x + 5 = 2(x + 5),$$

and opening brackets yields:

$$3x + 5 = 2x + 10.$$

This is an equation, the creature most commonly identified with algebra. An equation provides information by which we are supposed to find the identity of the "criminal", the unknown number.

Leaving systematic treatment of equations to later chapters, let us solve this particular equation by a simple observation. On the left hand side there is one more x than on the right. This is counterbalanced by the fact that the 10 on the right is larger by 5 than the 5 added on the left. So, $x = 5$. John is 5 years old, and David is 15, three times the age of John. In 5 years John will be 10, and David will be 20 years old, precisely 2 times John's age.

A diagram illustrates this nicely. In the following diagram, the top row is of length $3x + 5$, and the bottom is $2x + 10$. The equality between them shows that $x = 5$.

x	x	x	5
x	x	5	5

The first book on algebra was published by the Baghdad-ian mathematician Al-Houarismi, of the 12-th century. He named his book after the method for solving equations that we have just used, that of balancing the two sides. The word for balancing is "Al gabar", and this is the way algebra got its name.

Why is the Letter x Popular for Denoting Variables and Unknowns?

The letter x is by far the most popular letter in algebra. One theory explaining this is that the medieval Arabs did not use letters, but called the variable or unknown "thing", in Arabic "shei". The Spaniards replaced this by "chei", which is spelled in Spanish "xei". From the first letter of this word they borrowed the notation.

Substitution — connecting to the ground

A tip to be used throughout the study of algebra: the best way to understand an algebraic expression is to substitute numbers for the letters. For example to help a student solve an equation like $(a - 1) \times (a + 3) = 0$, prompt him or her to put numbers for a. A few attempts will lead him or her to the conclusion that one of the factors should be zero, which happens for $a = 1$ and for $a = -3$.

The Diophantus Riddle

The riddle on Diophantus' grave asks for the length of his life. Here it is:

> The youth of Diophantus lasted $\frac{1}{6}$ of his life. After another $\frac{1}{12}$ of his life he grew a beard. After another $\frac{1}{7}$ of his life he married. After 5 more years his son was born. His son died at half the age Diophantus died. Diophantus died 4 years after his son. How long did Diophantus live?

Diophantus

Diophantus, who lived in the third century AC in Alexandria, was the father of number theory and some claim also of algebra. He invented the use of formulas, and was the first using them to solve equations. His book "Aritmetica" was a main source of inspiration for the renaissance mathematicians. It was in the margins of a copy of this book that Pierre de Fermat wrote his famous conjecture, solved only in 1995, that for $n > 2$ there are no non-zero solution to the equation $x^n + y^n = z^n$. He added a comment that tormented mathematicians for over 300 years: "I have a marvelous proof, but there is not enough space in the margins to write it."

Solution: Denote Diophantus' life span (measured in years) by x. His youth lasted $\frac{1}{6}x$ years, he grew a beard at $\frac{1}{6}x + \frac{1}{12}x$, married at $\frac{1}{6}x + \frac{1}{12}x + \frac{1}{7}x$, and gave birth to his son at the age of $\frac{1}{6}x + \frac{1}{12}x + \frac{1}{7}x + 5$. The son died after $\frac{1}{2}x$ years, when Diophantus was $\frac{1}{6}x + \frac{1}{12}x + \frac{1}{7}x + 5 + \frac{1}{2}x$. Diophantus died after 4 years, namely when he was $\frac{1}{6}x + \frac{1}{12}x + \frac{1}{7}x + 5 + \frac{1}{2}x + 4$, and by the choice of x he was then x years old. We get:

$$\frac{1}{6}x + \frac{1}{12}x + \frac{1}{7}x + 5 + \frac{1}{2}x + 4 = x$$

and gathering terms we obtain:

$$\left(\frac{1}{6} + \frac{1}{12} + \frac{1}{7} + \frac{1}{2}\right)x + 5 + 4 = x$$

or $\left(\frac{14+7+12+42}{84}\right)x + 9 = x$, namely $\frac{75}{84}x + 9 = x$. Let us solve this by a guess (an easy one) — $x = 84$. Diophantus lived 84 years.

Part 2

Forward, or Backward to Elementary Mathematics?

How do I think about my problems? Concretely and systematically.

Karl Friedrich Gauss

Middle school teachers, all over the world, face the same dilemma: to repeat, or not to repeat? Should they devote the first month or two in the first year of middle school to rehearsing the primary school material, or start a new subject? My own unequivocal answer is — repeat. There is no choice. To understand what is $2x$, you first need to know the meaning of multiplication. For a student who does not understand the meaning of $\frac{3}{5}$, the expression $\frac{x}{x+2}$ is an empty shell.

To convince you of the necessity of going back to elementary school, here is a small challenge — four questions on elementary school mathematics, each hiding some insight that will be needed in algebra. Do they look simple?

1. What is addition?
2. Why is the fraction line a sign of division? Why $\frac{2}{3} = 2 : 3$?
3. Why is $\frac{2}{3} \times 24$ the same as $\frac{2}{3}$ of 24?
4. Can you tell an arithmetic story clarifying why $2 : \frac{1}{4} = 8$?

Here are the answers:

1. Every arithmetic operation has a meaning, a link to reality, telling us when is the operation applicable. The meaning of addition is joining: $2 + 3$ means joining 2 objects and 3 objects. The outcome is 5 objects.
2. Ask first: how will you divide an apple and a banana among 3 children? Of course, you divide each separately — divide the apple to 3, and then the banana to 3.

The darkened part is a third of apple + banana.

In the expression 2:3, one divides two equal objects, say two rectangles, to three portions. We have just learnt how to do it — divide each separately, as in the picture.

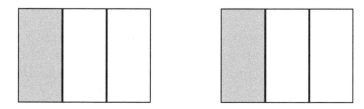

As seen from the picture, the result is two thirds. So, $2 : 3 = \frac{2}{3}$, which is what we wanted to show.

3. Multiplication is "times". You repeat the object. For example, 2 times an apple is 2 apples. This is also true for fractions. Multiplying something by $\frac{2}{3}$ means taking $\frac{2}{3}$ of this thing.

4. $6 : 2 = 3$ because when you divide 6 apples among 2 children each gets 3 apples, but also for another reason: 2 goes 3 times into 6. When you divide 6 apples among children and each gets 2 apples, there are 3 children. Similarly, $2 : \frac{1}{4}$ means "how many times $\frac{1}{4}$ goes into 2?" Since $\frac{1}{4}$ goes 4 times into 1, it goes $2 \times 4 = 8$ times into 2. A possible story: I divided 2 apples equally among children, and each got a quarter of an apple. How many children were there?

Repeating While Going Forward

Repeating elementary school material has its drawbacks. Its meaning for the students is "We are now going to not understand quickly what before we did not understand slowly". Indeed, the repetition is pointless if it is done in the same way as it was done before. Some new element is needed.

In fact, the innovation should be double. First, more stress is needed on the meaning of the operations, as opposed to their calculation. The other new element is introducing algebra at the same time. These two converge to the same solution: teaching the meaning of the arithmetical operations, using variables. When an expression contains variables, it cannot be calculated, and hence one concentrates on meaning. For example, $x + 2$ cannot be calculated without knowing what is the value of x, but one can understand its meaning.

The Meaning of the Arithmetic Operations

An arithmetic operation has two sides: calculation and meaning. Calculation answers the question "how much", while meaning tells you when is the operation needed. In algebra calculation is secondary, and hence we shall concentrate on the meaning of the operations.

Addition

The first meaning of addition is joining: in my right pocket there are 3 pencils, and in my left pocket there are 4 pencils. How many pencils do I have together? The answer is $4 + 3$. Note that we are not interested in the result of this exercise, only its form.

In the expression $4 + 3 = 7$ the numbers 3 and 4 are called "summands", and 7 is the "sum".

Precise terminology is a must. Words are the cement of thought. In order to stabilize some knowledge and make it available for future use it must be formulated in words. Once formulated explicitly, the next layer of knowledge can be built on it.

Every operation has more than one meaning. Addition has also the meaning of "more than":

Joseph's grade is 95, and Mary got 5 points more. What is Mary's grade? Write only the expression.

Algebraic Generalization

When the outcome of the expression is immaterial, it does not matter what numbers it includes. Instead of "Jane has 6 flowers and Ruth has 5 flowers" we can say "Jane has a flowers and Ruth has b flowers, how many do they have together?". The answer, $a + b$, is called "an algebraic expression", and a and b are called "variables".

Images

Every abstraction is accompanied in our mind by some image, though often it is vague and unconscious. A variable appears in our minds as some lump, representing a number. In the beginning of the study of algebra it is worthwhile to use images. For example:

> Ben has u dollars, and James has 5 dollars more. How many dollars do they have together?

Think of the sum owned by Ben as a rectangle.

James' $u + 5$ dollars can then be depicted as:

u	5

Joining the two gives $u + (u + 5)$, which is twice u plus 5, written as $2u + 5$.

Subtraction

In the most common meaning of subtraction, something disappears:

> Dirk had 10 pencils, 3 broke, how many intact pencils is he left with?

In a second meaning nothing disappears, but the objects are divided into types:

> Igor has 10 flowers, red and yellow, 3 of them are red, how many are yellow?

Yet a third meaning is comparison:

> Rebecca has 10 pencils, Dan has 3 pencils, how many more does Rebecca have than Dan?

Types of subtraction stories

Something disappeared: 2 of the 5 balloons popped, how many are left?

Partitioning to types: out of the 5 fruits 3 are apples and the rest bananas. How many bananas there are?

Comparison: by how many meters is the 324 meters Eiffel tower taller than a 187 meters tower?

In the subtraction exercise $10 - 3 = 7$, the number 10 is called "subtrahend", 3 is called "subtractor" and 7 is the "difference".

Multiplication

Multiplication means repeating the same thing a few times. For example, 3 times an apple means 3 apples. Yes, it is possible to multiply apples! It is not commonly known but counting and multiplication are one and the same thing. But usually we reserve the name "multiplication" for repeating numbers. 3 times 4 means repeating 4 three times, namely $4 + 4 + 4$. Note that this is not the same as 4 times 3, which is $3 + 3 + 3 + 3$. The result, 12, is the same, but not the expressions.

Why 2 Times x is 2x?

The ancient Indians (the Asian ones) wrote "3 4" for "3 times 4". We no longer use this notation, for the risk of confusing it with the number "thirty four", but there is logic in it. "Two apples" we may write this way:

For the same reason 2 times x, namely $2 \times x$, can be written as $2x$.

The Notation for Multiplication

The notation "$\times$" for multiplication was suggested by the Englishman Willliam Ottred in 1631. At about the same time the use of the letter x in algebra became common, and the danger of confusion between the two arose. To overcome the problem, the German mathematician and philosopher Gottfried Leibnitz suggested the use of a dot instead: $2 \cdot 3$. Now the expression $2 \cdot x$ can be written without risk of ambiguity. But for the reason pointed out above, the dot is not really necessary, and nowadays the most common notation is $2x$.

The Commutativity of Multiplication

2×30 means 2 times 30, which is $30 + 30$, while 30×2 means 30 times 2, which is $2 + 2 + \cdots + 2$ thirty times. Why are they equal? Suppose that there are 30 students in class. Each has 2 feet, and so together they have $2 + 2 + \cdots + 2 = 30 \times 2$ feet. Now let us count the feet in a different way: there are 30 right feet, and 30 left feet, together $30 + 30 = 2 \times 30$ feet. Of course, the two ways should lead to the same result.

The law of commutativity, $a \times b = b \times a$, is not an axiom, namely it did not "fall from heaven". It is a theorem that can, and should, be proved.

In one pack of bottles there are 6 bottles. How many are there in x packs?

Answer: $x \times 6$.

Write a story for which the right expression is $6 \times x$.

When interested in the result, and not in the meaning, we shall use the law of commutativity freely. We shall write $6x$ for $x \times 6$, and actually prefer the first to the second.

Daria has a bills of \$20, and b bills of \$50. What is the expression for the total amount she has? What are a and b if the total amount is \$200? Is there more than one possible answer?

(Answer: the expression is $20a + 50b$, and there are 3 possibilities for this amount to be \$200: $a = 0$, $b = 4$; $a = 5$, $b = 2$; $a = 10$, $b = 0$.)

Division

In division, a set or an abstract number is divided into equal parts. There are two questions that can be asked: what is the size of each part, and how many parts there are. So, there are two principal meanings to division.

I put 12 flowers in 3 vases, the same number in each. How many flowers are there in each vase? What type of division is this?

Of course — "how large is each part", and the expression is $12 : 3$.

A father divided 12 candy bars among his children, and each got 3 bars. How many children were there?

Again, the answer is $12 : 3$, but this time the division is of the type "how many parts". In the first story $12 : 3 = 4$ because $4 + 4 + 4 = 12$ (4 flowers in each of the 3 vases), namely $3 \times 4 = 12$; while in the second $12 : 3 = 4$ because $3 + 3 + 3 + 3 = 12$, namely $4 \times 3 = 12$. So, the existence of two different meanings to division stems from the commutativity of multiplication.

A ship brought a cargo of x cars, 10 in each container. How many containers were there? What type of division is this?

(Solution: $x : 10$, the type is "how many parts".)

In a chocolate bar there are 20 squares, and each row contains r squares. How many rows there are? What type of division is this?

Discrete and Continuous Quantities

I want to make here an important distinction, that appears often but is rarely mentioned explicitly, between discrete and continuous quantities. Discrete objects are those that we can count, like apples. But how do we "count" length, or air pressure, or temperature, or the amount of electricity you use in a month? These are continuous magnitudes, and they cannot be counted. To measure them, arbitrary units of measurement are invented. For example, "foot" (which was probably indeed the length of an average person's foot) to measure length, or one percent of the distance in temperature between boiling and freezing water, to measure heat.

A major step forward was the standardization of measurements. In 1791, amidst their revolution, the French decided to standardize the measurement of length, and invented a unit of length called "meter", which they defined to be 1:10,000,000 of the distance between the equator and the north pole. Needless to say, the measurements of those times were not precise in our standards, and the French were about 0.5% off in measuring this distance. Nowadays a "meter" is defined in a totally different way: it is the distance light traverses in $\frac{1}{299,792,458}$ of a second.

By the way, can you guess how is a "second" defined?

Fractions

A Fraction Means Two Compounded Operations

I will dwell more on fractions than on any other subject of elementary mathematics, since fractions are the stumbling block for most middle school students, and most of the revision time should be devoted to them.

What is $\frac{5}{8}$ of 240 apples? One eighth of 240 is 240 divided by 8, namely 30, and $\frac{5}{8}$ is just what you hear when you say the word — 5 eighths, namely 5 times 30, which gives 150.

A fraction is the conjunction of two operations, division and multiplication. $\frac{5}{8}$ of an object, where the object can also be a number or a set, is obtained by dividing it by 8, and multiplying the result by 5.

> The conjunction of addition and subtraction is not a new operation: subtracting 8 and adding 3 is just like subtracting 5. The conjunction of division and multiplication is something new. It demands the invention of a new concept — the fraction.

An Activity for Understanding Fractions

In every class in which I teach fractions I do one activity that in my view is a must. I draw on the board some number, say 10, of objects — say circles. I ask the students to draw in their books a similar picture, and circle one fifth, namely 10:5 (for the sake of the example) of the objects. I circle two objects on the board, and write "one fifth of 10 is 2", and also "$\frac{1}{5}$ of 10 is 2". I then ask them to circle another fifth, and we write "$\frac{2}{5}$ of 10 are 4". We then circle another fifth, and another, and write the corresponding words. When we get to 5 fifths of 10, which is of course 10, I ask them why this is so, and what are $\frac{5}{5}$ of a million — they are proud of doing calculations in the millions. Then I ask them what are $\frac{6}{5}$ of 10. This cannot be drawn any more because there are only 10 circles on the board, but we can calculate: one fifth is 2, so 6 fifths is 6 times 2, namely 12. We go on with this, up to (say) 20 fifths.

> Draw 8 flowers, and circle a quarter of them. How much is a quarter of 8? Circle another quarter. How many flowers are two quarters of 8 flowers? Continue this way to 4 quarters. How much is 100 quarters of 8? What are the operations you used to calculate the answer?

In the notation "$\frac{3}{4}$" the 4 is called "denominator" because it gives the name to the fraction, and 3 is the "numerator" because it enumerates (counts) how many parts are taken.

> What is one eighth of 24? What are *a* eighths of 24?

Answer: $3a$.

The Whole

The object from which the fraction is taken is called the "whole". It can be an object, a set, a shape or a number. The whole serves the same role as denomination in counting: in "$\frac{3}{4}$ of an apple" the whole is the apple, just as in "3 apples" the denomination is "apple".

Teaching Division and Fractions Together

Unfortunately, in elementary school fractions and division are studied separately. Division is studied in Grade 2, while fractions are studied in Grades 3 or 4. This is a mistake: the two should be taught together. In Grade 1, when division by 2 is studied, the result should be given a name — half, and the notation $\frac{1}{2}$ should be introduced. In Grade 2, when division by 4 is studied, the notion of quarter can be mentioned, and also what are three quarters. Division and fractions are almost the same, and teaching them together will make fractions much more accessible for the children.

A main factor in the separation of the two subjects is that division is studied only of numbers, while fractions are first taken of shapes, mainly a circle. This is yet another mistake. Division should also be of shapes — what is a square divided by 3, or a rectangle divided by 5. And fractions should be taken, right from the start, also from numbers — what is a half of 12 circles? $\frac{3}{4}$ of 12 apples?

Another Way of Understanding What $\frac{3}{4}$ is

— *How do you divide 3 apples among 4 children?*
— *You make apple mash.*

This is time to learn why the fraction line can be viewed as division. How do you really divide the apples? Repeating the same idea of dividing an orange and a banana among three children from the introduction, let us divide 3 cakes (rectangles) among 4 children. Here is the picture — we divide each rectangle separately.

Each child got 3 quarters. We learn that $3 : 4 = \frac{3}{4}$, which explains why the fraction line is sometimes called also "division line". A fraction is the result of dividing its numerator by its denominator.

Two Rules on Fractions

What Happens When the Numerator is Multiplied by a Number?

There are two rules that put order in the world of fractions. One is:

Multiplying the numerator of a fraction by (say) 4 multiplies the fraction by 4.

For example, if 5 pirates share 3 gold bars, each gets $\frac{3}{5}$ of a bar. If the number of the bars turns out for some reason to be 4 times bigger, namely 12 bars, the share of each grows 4 times. That is, $\frac{12}{5} = 4 \times \frac{3}{5}$.

This tells us that to calculate $4 \times \frac{3}{5}$ we should multiply the numerator by 4.

Calculate $2 \times \frac{3}{5}$.

Solution: $\frac{2 \times 3}{5} = \frac{6}{5}$.

Calculate $m \times \frac{2}{5}$.

Solution: $\frac{2m}{5}$.

Generalize: what is $m \times \frac{a}{b}$?

What Happens When the Denominator is Multiplied by a Number?

The second rule is:

When the denominator is multiplied by 4, the fraction is divided by 4.

This is also clear: if there are 4 times more pirates, the share of each is 4 times smaller. For example, if instead of 3 pirates sharing the 5 bars we have 20 pirates, the share is $\frac{3}{20}$, and it is 4 times smaller, namely $\frac{3}{20} = \frac{3}{5} : 4$.

This is used for dividing a fraction by a number: $\frac{3}{5} : 4$ is obtained by multiplying the dominator by 4, to obtain $\frac{3}{20}$.

Of course, 4 is just an example. The rule is:

Multiplying the numerator by a multiplies the fraction by a. Multiplying the denominator by a divides the fraction by a.

What fraction in each pair is smaller, and by what factor?
a. $\frac{1}{4}, \frac{1}{12}$ b. $\frac{1}{8}, \frac{1}{40}$ c. $\frac{3}{8}, \frac{3}{40}$ d. $\frac{2}{15}, \frac{2a}{15}$ e. $\frac{2}{15}, \frac{4a}{15}$ (*) f. $\frac{4a}{3}, \frac{7a}{3}$.

Calculate: a. $\frac{2}{3} : 4$ b. $\frac{2}{3} : 5$ c. $\frac{2}{3} : m$ d. $\frac{a}{b} : m$.

(Solution for d: $\frac{a}{bm}$.)

Multiplying Two Fractions

Using the two rules, we now know how to multiply two fractions. Take for example the product $\frac{3}{4} \times \frac{2}{5}$. We know that $\frac{3}{4} = 3 : 4$. So, multiplying by $\frac{3}{4}$ means multiplying by 3:4, which means multiplying by 3 and dividing by 4. By the two rules we learnt, multiplying $\frac{2}{5}$ by 3 means multiplying it numerator by 3, and dividing the result by 4 means multiplying the denominator by 4. We get $\frac{3}{4} \times \frac{2}{5} = \frac{3 \times 2}{4 \times 5}$. In general:

Multiplying two fractions is done by multiplying numerators to get the numerator of the result, and multiplying denominators to get the denominator of the result.

Algebraically:

$$\frac{a}{b} \times \frac{c}{d} = \frac{ac}{bd}.$$

Calculate

$$\frac{1}{2} \times \frac{1}{3}, \frac{1}{4} \times \frac{1}{5}, \frac{1}{m} \times \frac{1}{n}.$$

Calculate:

$$\frac{5}{3} \times \frac{3}{5}, \frac{1}{2} \times \frac{2}{1}, \frac{m}{5} \times \frac{5}{7}, \frac{a}{b} \times \frac{b}{a}.$$

Expansion and Reduction of Fractions

Doing an Operation and Then Reversing it

Alice put her hand on her head, to see if she is growing or getting smaller. To her surprise, she felt no change. Of course, this is what usually happens when one eats a cake.

'Alice in Wonderland', Lewis Carrol

Eating from one side of the mushroom, Alice grew four times her previous height. Eating from the other side, she became 4 times smaller, thus getting back to her original size. Just like if you walk 4 kilometers south and then 4 kilometers north, you get back to where you started.

By our two rules, if we multiply the numerator of a fraction by 4 it grows 4 times; if we then multiply the denominator by 4, it diminishes 4 times, getting back to its original size. For example, if we multiply the numerator and denominator of $\frac{2}{5}$ by 4, we get $\frac{4 \times 2}{4 \times 5} = \frac{8}{20}$. So, $\frac{8}{20} = \frac{2}{5}$. What we just did is called *expanding* the fraction $\frac{2}{5}$ by 4. Expanding a fraction by a number means multiplying its numerator and denominator by this number. Expanding the fraction does not change is value.

Expand the fraction $\frac{2}{5}$ by *a*.

Solution: $\frac{2}{5} = \frac{2a}{5a}$.

Expand the following fractions by *x*: a. $\frac{1}{2}$ b. $\frac{2}{3}$ c. $\frac{a}{b}$.

Solution of c: $\frac{ax}{bx}$.

Reduction

The opposite of expansion is called reduction. It means dividing the numerator and denominator by the same number. Reduction, like expansion, does not change the value of the fraction. For example, reducing $\frac{4}{6}$ by 2 yields: $\frac{4}{6} = \frac{4:2}{6:2} = \frac{2}{3}$.

> Reduce the following fractions, and write by which number you reduced them: a. $\frac{3}{3}$ b. $\frac{4}{2}$ c. $\frac{6}{12}$.

A fraction that cannot be reduced by any number larger than 1 is called a "reduced fraction".

> Which of the following fractions is reduced? $\frac{2}{8}$, $\frac{2}{7}$, $\frac{21}{90}$.

We can also reduce algebraic fractions. For example, $\frac{a}{2a}$ can be reduced by a, to yield $\frac{1}{2}$.

> Reduce the following fractions: $\frac{a}{a}$, $\frac{2a}{a}$, $\frac{2x}{3x}$, $\frac{ax}{a}$.

> Complete: a. $\frac{1}{2} = \frac{}{4}$ b. $\frac{1}{2} = \frac{3}{}$ c. $\frac{a}{b} = \frac{2a}{}$, $\frac{2a}{3b} = \frac{}{9b}$.

Division of Fractions

Invert and Multiply

Dividing by a fraction is the opposite of multiplying by it. So, if multiplying by $\frac{3}{4}$ is done by multiplying by 3 and dividing by 4, dividing by $\frac{3}{4}$ is done in the reverse way — dividing by 3, and multiplying by 4, which really means multiplying by $\frac{4}{3}$. The general rule is: dividing by $\frac{m}{n}$ is multiplying by $\frac{n}{m}$. American teachers who despaired of explaining to their students why this is so, composed a rhyme (after a famous poem by Alfred Tennyson):

> Ours not to reason why
>
> Just invert, and multiply.

Here is a conversation I had with my then 10 years old daughter, Geffen, in which I taught her the rule for division. I am bringing it mainly as an exemplifier of a few teaching principles.

A Conversation on Division by $\frac{2}{3}$

I: Look at the exercise $10 : \frac{2}{3}$. How do you think we should start?

G: We should ask a simpler question. (**It was Geffen who taught me this principle: when asked a question that was too hard for her, she would ask for a simpler question.**)

I: Indeed. What is the simplest fraction you can think of?

G: $\frac{1}{2}$.

I: Indeed, $\frac{1}{2}$ is a simple fraction. Can you invent a simple division exercise?

(Start from the simplest possible example).

G: $1 : \frac{1}{2}$.

I: and how much this is?

G: Yes, this is 2, because $1 : 2 = \frac{1}{2}$. If $6 : 2 = 3$ then $6 : 3 = 2$. If $1 : 2 = \frac{1}{2}$ then $1 : \frac{1}{2} = 2$.

I: Very nice. But here is another way. $6 : 2 = 3$ because 2 goes 3 times into 6. Remember what kind of division is this?

G: Yes, this is "how many parts".

I: How many times does $\frac{1}{2}$ go into 1?

G: 2.

I: so what is $1 : \frac{1}{2}$?

G: 2.

(Teach the same principles from different angles. We learnt two ways of calculating $1 : \frac{1}{2}$.)

I: Now can you tell me what is $3 : \frac{1}{2}$? **(Introduce one principle a time.)**

G: 6, because $\frac{1}{2}$ goes into 1 twice, so into 3 it goes $3 \times 2 = 6$ times.

I: Nice. And what is $4 : \frac{1}{2}$? **(Stabilize the principle by exercising)**

G: 8.

I: What is the rule?

G: Dividing by $\frac{1}{2}$ is multiplying by 2. **(We reached the rule by examples, now it is time for putting it into words.)**

How many halves of apples there are in 10 apples? What is $10 : \frac{1}{2}$?

I: Let us now move to division by $\frac{1}{3}$. What is $1 : \frac{1}{3}$?

G: $\frac{1}{3}$ goes 3 times into 1, so $1 : \frac{1}{3} = 3$.

I: And $4 : \frac{1}{3}$?

G: In 4 the fraction $\frac{1}{3}$ goes 4 times 3 times, so $4 : \frac{1}{3} = 12$.

I: Can you tell me the rule?

G: Yes, dividing by $\frac{1}{3}$ is multiplying by 3.

I: And dividing by $\frac{1}{4}$?

G: It is multiplying by 4.

In the language of algebra, the rule is: dividing by $\frac{1}{n}$ is multiplying by n.

I: Now let us divide by $\frac{2}{3}$. If we divided 10 cakes among children, and each child got $\frac{1}{3}$ of a cake, how many children were there?

G: $10 : \frac{1}{3} = 30$, there were 30 children.

I: Now suppose that every child got $\frac{2}{3}$ of a cake, namely each child got 2 times as before, how many children there are now, fewer or more?

(I did not ask directly "how many times fewer", just "fewer or more", letting her ask the questions "how many times" by herself.)

G: 2 times fewer, because each child gets a portion that sufficed before for 2 children. So, there were 15 children.

I: so, what is $10 : \frac{2}{3}$?

G: 15.

I: How did we get to this answer?

G: We multiplied by 3, and then divided by 2.

I: Can you tell me the rule, what happens when you divide by a fraction?

G: Yes, you divide by the numerator, and multiply by the denominator.

Example: $24 : \frac{3}{4} = (24 : 3) \times 4 = 8 \times 4 = 32$.

Calculate: $8 : \frac{1}{3}$, $8 : \frac{2}{3}$, $10 : \frac{5}{4}$, $\frac{p}{q} : \frac{p}{q}$, $\frac{p}{q} : p$.

Part 3

Negative Numbers

Reality or Imagination?

Until about 200 years ago, even professional mathematicians were suspicious of negative numbers. Carnot, a member of the French academy, wrote in 1803 that "a negative number is obtained by deducting from a non-existing entity, a plain absurdity". Toward the end of the 19-th century a French educationalist blamed the problems of French education on the teaching of negative numbers. August De Morgan, a well known English mathematician, claimed that if a negative number appears as an answer to a question, it only means that we did not ask the right question.

For example, consider the following problem.

John is 8 years old, and his brother Dick is 5. In how many years will John be twice as old as Dick?

The answer is (-2) (the parentheses are there to clarify that this is a minus sign, and not a dash). In (-2) years, John will be (that is, was) 6, twice the age of Dick, who was 3. Many of today's children are familiar with the idea that "in (-2) years time" means "2 years ago". De Morgan claims that we did not formulate the problem correctly. We should have asked "how long ago was John twice the age of Dick?"

How to solve De Morgan's problem? Ask a simpler question. Deborah is 2 years old, and her sister Anabel has just been born. When will Deborah be twice the age of Anabel? Of course, in 2 years time, when Deborah

(Continued)

29

(Continued)

is $2 + 2$ and Anabel is 2. In general, when person A is k years older than B, when B is k years old A is $2k$ years old, twice the age of B. In our case, the difference in ages between John and Dick is 3, so when Dick was 3 John was 6, twice Dick's age. And this happened 2 years ago, or in terms of negative numbers, in (-2) years' time.

Here is another age problem with a negative answer:

Ben is 15, and his father 39. When will the father be 3 times older than Ben?

Solution: When this happens, the difference between their ages will be 2 times Ben's age (see illustration). Since the difference is 24 years, this "will" happen when Ben is 12, namely 3 years ago, namely in (-3) years time.

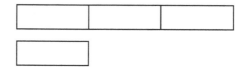

The use of negative numbers spread in Europe only in the 14-th century. In India this happened 700 years earlier. For the mathematician of today, negative numbers are no less real than positive numbers. They denote real things. But the difficulty with which negative numbers were accepted shows that their understanding demands abstraction.

Positive and negative numbers are called together "directed numbers". The name comes from the representation on the real line: positive numbers are directed right, negative numbers are directed left.

Introducing Negative Numbers by Sequences

A good way of introducing negative numbers is through sequences. Sequences whet the appetite for finding regularity. Look at the sequence 5, 4, 3, 2, 1, and try to continue it. The next element is of course 0. Even in first grade you will find children who know the next number, (-1) (as already noted, the parentheses clarify that this is not a dash). This is 1 below zero. Next comes 2 below zero, which is (-2), and then (-3).

Now try to continue the sequence $7, 5, 3, 1$. Since in every step the elements decrease by 2, the next element is (-1), to be followed by $-3, -5, -7$.

Continue the sequences: a. $20, 15, 10, \_, \_, \_, \_,$ b. $13, 8, 3, \_, \_, \_, \_$
c. $50, 40, 30, \_, \_, \_, \_, \_$ d. $500, 400, 300, \_, \_, \_.$

The numbers $1, 2, 3, 4, \ldots$ are called "positive", the numbers $-1, -2,$ $-3, -4, \ldots$ "negative". Zero is neither positive nor negative.

So as not to discriminate the positive numbers, they are also given a sign, "+". Instead of 4 we may write "+4".

Opposite Numbers

10 and (-10) are called "opposite". Generally, the opposite of a number is obtained by reversing the sign. The opposite of a number x is denoted by $(-x)$. For example, $(-(-3))$ is the opposite of (-3), which is 3.

The opposite of the opposite of a number is the number itself. The opposite of a number a is also called "minus a".

Is there a number that is the opposite of itself?
What is the opposite of the opposite of the opposite of (-5)?

The Three Different Meanings of the "−" Sign

The "−" sign has three different meanings. In the expression $4 - 3$ it denotes the binary (two numbers) operation of subtraction. In (-3) it denotes direction, and in $(-(-3))$ the left sign denotes "opposite", namely it reverses the sign of (-3). In the last meaning, it is an operation on one number, a "unary" operation. This is a bit confusing, and indeed there are suggestions of changing the notation, writing $\overline{3}$ for (-3). But it is hard to change a 600 years old notation, and humanity is probably stuck with this notation for a while.

The minus sign works hard. It is used to denote three different things:

a. subtraction.
b. negative numbers
c. the opposite of a number.

This is not totally illogical. The three meanings are closely related.

Minus Minus is Plus

In a thank you letter of 7-th graders at the end of the academic year, a student wrote "we did not not enjoy your classes". "Not not" is "yes", and likewise "minus minus" is plus. $-(-a) = a$.

> Genny said: I did not not not bring a sandwich to school today". Did she bring a sandwich or not?

An odd number of no's is "no", and an even number of no's is a yes. Likewise, an odd number of minuses is a minus, and an even number of minuses is a plus.

> What is "yes yes"? What is $+(+a)$?

A mathematician explains to his audience: "Negation of negation is affirmation. Negation of affirmation is usually negation. But affirmation of affirmation is without exception affirmation."

Derisive call from the audience: "Yeah, yeah."

> What is $-(-(-(-3)))$?

Meanings of Directed Numbers

Debts

To really understand negative numbers, we need to give them meaning, namely connect them to reality. The first meaning historically used, by the ancient Indians, was that of debts. If somebody is penniless, and in addition he owes $100, he has (-100). Owning a debt of x dollars means having $(-x)$ dollars.

A debt of x dollars means having x dollars less than 0, namely having $(-x)$ dollars.

Floors, and a First Encounter with Operations with Negative Numbers

In Europe the ground floor is numbered 0. In the US it is numbered 1. We shall stick here with the European system. The underground floors are numbered (-1), (-2) and so on.

Using this meaning, we can have first acquaintance with operations on negative numbers. Floor (-8) is 8 floors below ground. Descending one floor brings you to floor number (-9), hence $-8 - 1 = -9$. Likewise, the floor above (-8) is (-7), so $-8 + 1 = -7$.

i. What is the number which is 2 less than (-8)? What is $(-8) - 2$?
ii. What is the number larger by 2 than (-8)? What is $(-8) + 2$?
iii. Add 1 to each of the following numbers: 100, (-100), 1000, (-1000), 6789, -6789.
iv. Daria stands on the 4-th floor. She wishes to get to the floor numbered (-2) (2 floors underground.) How many floors should she go down? Conclude what is $4 - (-2)$.
v. Danna stands on the 5-th floor. How many floors should she go *up* to get to the 3-rd floor?

i. (Solution: (-10).)
ii. (Solution: (-6).)
iv. (Hint: do it in two steps. First let her go to floor 0 — this takes 4 floors to go down. Then ask her to descend 2 more floors, to floor (-2). Together 6 floors. $4 - (-2) = 6$.)
v. (Solution: (-2).)

Above and Below Sea Level

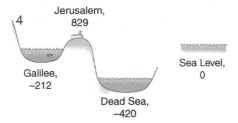

Israelis and Jordanians have an advantage over other nations in the study of negative numbers: they can experience in real life what is negative height, measured with respect to sea level. Sea level is arbitrarily (though quite naturally) set to be zero, and altitudes that are below it are denoted by negative numbers. Jointly in Israel and in Jordan there is the lowest point on earth, the Dead Sea, which is 420 meters below sea level, and the recession of the sea makes it lower about 1 meter every year.

Let us use this meaning of negative numbers to exemplify subtraction resulting in a negative number, namely subtraction of a big number from a smaller one.

A car starts at altitude 500 meters, and goes 800 meters down. What altitude does it reach?

Solution: Do it in two steps. First let the car go 500 meters down, reaching height 0 (namely, sea level). Then it goes the remaining 300 meters, which brings it to altitude (-300). The exercise is written: $500 - 800 = -300$.

A hot air balloon starts at altitude 2000, meters, and drops 2400 meters. What altitude did it reach? Where did it land?

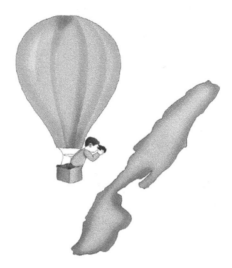

The tallest mountain in Jordan is 1800 meters high. How much should one go down to reach the Dead Sea from its top? Write it as a subtraction exercise.

(Solution: $1800 - (-420) = 2420$ meters.)

Temperature

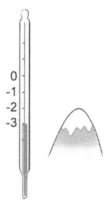

In continental Europe, as in most non Anglo-Saxon countries, a scale of temperatures is used that was invented by the Swedish physicist Celsius (1701–1744). In this scale, the freezing temperature of water at sea level is defined to be 0 degrees (the freezing temperature, and more so the boiling temperature, depends on air pressure, and so on the height above sea level.) The boiling temperature was set to be 100 degrees. The scale is then spread evenly (in some sense that can be made precise) between the two temperatures and beyond. A curiosity: Celsius himself chose reversely, the boiling temperature as 0 and the freezing temperature as 100. Had humanity stuck to this convention, the temperature in the Antarctic would be said to be "high". Perhaps out of respect to Celsius, the direction was reversed only a year after his death.

> The temperature was 3 below zero, and dropped 7 degrees. What is the temperature now?

(Answer: $(-3) - 7 = -10$.)

At around the same period of Celsius, a Dutch physicist named Fahrenheit was trying to invent a temperature scale that would avoid negative temperatures. He chose as 0 the lowest temperature reachable in his time. That was a sign of sheer lack of imagination. Not long after, much lower temperatures were reached. In spite of its unnaturalness, the Americans chose it as their official scale.

Where Fahrenheit failed, another physicist, Lord Kelvin, succeeded. He discovered that there is a lowest temperature, an "absolute zero". This is not so surprising: if temperature corresponds to motion of molecules, zero motion means a lowest temperature possible. There is absolute motionlessness. The lowest temperature is about (-273) on the Celsius scale. It is also called "0 Kelvin". The Kelvin scale is the Celsius scale, shifted. A shift of one degree Kelvin is a shift of one degree Celsius. So, (-272) Celsius is 1 on the Kelvin scale.

To economize notation, Celsius degrees are denoted by a small circle, plus C. For example, 30 degrees Celsius is denoted by $30°C$. Kelvin degrees are denoted by K, without the circle. 30 degrees Kelvin is denoted by $30K$.

> Calculate the following in Kelvin degrees: $0°C, 1°C, 2°C, 3°C, x°C$.

(Partial solution: $0°C = 273K$, $2°C = 275K$, $x°C = (273 + x)K$.)

> Write in the Celsius scale: a. $0K$ b. $273K$ c. $10K$ d. xK.

(Solutions: a. $(-273)°C$ b. $0°C$ c. $(-263)°C$ d. $(x - 273)°C$.)

Operations with Directed Numbers

Addition of Negative Numbers

What is $(-5) + (-3)$? If Reuben owes person A 5 dollars, and person B 3 dollars, together he owes 8 dollars. So, $(-5) + (-3) = (-8)$.

In general, $(-a) + (-b) = -(a + b)$.

This is called "distribuitvity". Minus the sum is the sum of the minuses.

Calculate: a. $(-1) + (-1) + (-1)$ b. $(-a) + (-b) + (-c)$ c. $(-3) + (-3)$ d. $(-3) + 3$ e. $-a - (-a)$.

Adding $(-b)$ Means Subtracting b

If your boss announces that you got a salary raise, check first whether the addition is positive. An addition of (-1000) dollars to the salary means a decrease of 1000 dollars. Here is another story exemplifying this:

Alex has 10 dollars, and he owes 3. How much money does he really have?

The answer is $10 + (-3)$, which is 7. When he pays his debt, he will be left with 7 dollars.

Adding a debt of 3 dollars means subtracting 3 dollars from your sum. Adding (-3) means subtracting 3.
In general: adding $(-b)$ means subtracting b.

Rebecca has \$30, and she owes Abe \$30. How much money does she really have? What is $a + (-a)$?

Subtracting a Number from a Smaller One

Negative numbers enable subtracting a number from a smaller number. We have already encountered this, in altitudes and temperatures. For example, $0 - 10 = (-10)$, since 10 units below 0 is (-10).

How does one calculate $3 - 13$, for example? By subtracting 13 in two steps. First subtract 3, to reach 0, and then subtract the remaining $13 - 3$, namely 10, that have not been subtracted as yet. We get $0 - 10$, namely (-10). This showed that $3 - 13 = -(13 - 3)$, and in general:

$$a - b = -(b - a)$$

The equality $a - b = -(b - a)$ is logical. Each of the two numbers appears in both sides with the same sign. b appears in both with the sign "–". a appears on the left in the positive, and on the right with a "minus minus" sign (minus outside the parentheses, and minus inside), which is also positive.

John had 10 dollars. He owes Rob some amount, and he calculated that he has (-20) dollars. How much does he owe Rob?

(Answer: 30 dollars, since $10 - 30 = (-20)$.)

Joe has $7, and he has to pay $20 for the annual trip. He gives the $7 he has. How much does he still owe?

(Answer: 20–7, showing that $7 - 20$, the amount he has, is $-(20 - 7) = -13$, which is the amount he owes after being left with 0 dollars.)

Complete the missing number: $6 - b = -(\_\_ - 6)$.

Subtracting a Negative Number

"Minus minus" is plus, and hence $10 - (-3) = 10 + 3$. This is logical: subtracting 3 dollars from your debt is tantamount to adding 3 dollars to your account. In fact, this is the same as a rule already formulated:

Subtracting a number is the same as adding its opposite.

Calculate a. $-5 - (-5)$ b. $10 - (-5)$ c. $5 - (-5)$ d. $2 - (-(2 - (-2)))$
What is $100 - (100 + 1)$? Generalize: what is $100 - (100 + x)$?

(Solution: $-x$.)

Calculate $100 - (100 - 1)$. What is $100 - (100 - x)$?

Minus Before Parentheses

The rule is: a minus before the parentheses reverses all signs inside the parentheses. Plus becomes a minus, and minus becomes a plus. For example: $-((-3) + 4 - 5 + (-8)) = 3 - 4 + 5 + 8$. Another example (algebraic, this time): $-(a - b) = -a + b$.

Open parentheses in $a - (b - c)$ and in $a - (b + c)$.
Calculate a. $0 - (-3)$ b. $1000 - (-3)$ c. $10 - (-10 - (-10))$ d. $(-17) - (-1)$.

The Years BC

Like with temperatures, to measure time it is necessary to fix an arbitrary time point which is set as 0. The Jews set this point at the creation of the world, which they believe to have occurred in the year the Christians count as (-3760). A choice unfriendly to mathematics, since there is no point in speaking about negative years, nothing happened then. The Christian counting is more useful in the study of negative numbers. They set as 0 the year of the birth of Christ, and so historical events did happen in years that are negative in this counting.

What was earlier, 320 BC or 319 BC? By how much time?

Negative years are a good opportunity to apply our rules. Between year x and year y there pass $y - x$ years. For example, between 2000 and 2004 there passed 2004-2000, namely 4 years. Hence the length of a person's life is the year of his death minus the year of his birth. This is only approximate, for example Mozart was born in 1756 and died in 1791, but lived almost 36 years, since he was born in January 1756 and died in December 1791.

Socrates, the Greek philosopher, was born in the year (-470), and was executed by his city mates, who resented his prodding questions, in (-399). He thus lived $(-399) - (-470) = 470 - 399 = 71$ years.

$(-399)-(-470) = 71$

Alexander the great was born 356 BC, and died at the age of 33. So, he died in the year $(-356) + 33$, which we know to be $-(356 - 33) = -323$.

Alexander established the city Alexandria in Egypt in (-328). How old was he then? Answer: $(-328) - (-356)$, which is $356 - 328$, namely 28. He was 28 at the time.

Calculate the life span of each of the following:

Archimedes, a Greek mathematician who lived in Sicily and calculated the volume of the ball:

$(-287)-(-212)$

Pythagoras, another Greek mathematician who lived in today's Italy.

$(-580)-(-500)$

Augustus, the first Roman emperor:

$(-63)-(14)$

The Order Among Directed Numbers

Now he is being punished, sitting in jail. The trial is going to start on
Wednesday next week, and last, of course, comes the crime.
                                "Through the Looking Glass", Lewis Carrol

Joseph owes 100 dollars, and Renata owes 101 dollars. Assuming that they
have no other assets except for their debts, who is richer? Of course, Joseph,
by 1 dollar. Meaning that $-100 > -101$. Likewise, the 7-th floor below
ground is lower than the 5-th floor below ground. So, (-7) is less than (-5).

In Lewis Carrol's "Through the looking glass" everything is reversed.
To go somewhere one needs to go in the opposite direction, and the order
of time is reversed. The White Queen utters a cry of pain, and then she is
pricked by a needle. Negative numbers are similar. They are a mirror world.
In the ordinary numbers $101 > 100$, and in the negative numbers the order
is reversed, $-100 > -101$.

Why is it that the mirror reverses right and left, and not top and bottom?

Answer: The mirror does not reverse anything. It reflects everything as it
is. We just believe it reverses right and left. This stems from the way we
are accustomed to face each other, face to face, which means right hand
opposite left hand and left hand opposite right hand. Because of this we
believe that the "person" in the mirror, like real people facing us, has his
right hand opposite our left, though in fact opposite our left hand is our
left hand, just as it should be.

If anywhere in the universe there are creatures whose mouths are in their
heads and their ears are in their feet, in order to communicate they stand
face to feet. They would believe that the mirror reverses up and down.

The lowest point on earth, as already mentioned, is the Dead Sea, with
altitude (-420) meters. The next lowest point (that is, one far from the
Dead Sea), is Turpan Valley in China, with (-158) meters altitude. So,
$-420 < -158$.

$$\text{If } a > b \text{ then } (-a) < (-b).$$

Order, from left to right, the numbers:

$$0, \ -5, \ -1, \quad -\frac{1}{2}, \ -\frac{2}{3}, \ (-10000).$$

The Directed Numbers on the Real Line

The directed numbers got their name from their depiction on the real line. In order to represent negative numbers, the line is continued to the left.

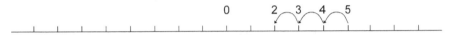

On the real line it is easy to visualize addition and subtraction. For example, adding 3 is done by going 3 units to the right, and subtracting 3 by going 3 units to the left. Here is for example the calculation of $5 - 3$:

Use the real line to calculate $0 - 3$, $-10 - 3$, $-4 - 3$, $-1 - 3$, $-1 - 3$.

Opposite numbers have a simple geometric meaning on the line. Starting from 0, going 5 units to the right takes you to $+5$, and going 5 units to the left takes you to (-5). The opposite of a number is its mirror image with respect to 0.

Where do you land if you start at point x and go $3x$ units left?

Absolute Value

Imagine a person who gets a note from the ATM that his balance is $\$(-10,000)$ and he is happy. He cares only for the size of the amount, not whether it is in the plus or in the minus. In mathematics there is such a "person". Its name is "absolute value". The absolute value of a number is its size, ignoring the sign. The absolute value of x is denoted by $|x|$. So, $|+7| = 7$, but also $|-7| = 7$, and $|-10,000| = 10,000$. The absolute value of a number is always non-negative, that is, positive or zero.

In some situations the sign is indeed not important. For example, if a person cares only how far did he go, not minding in which direction, it is the absolute value of the distance he went that is at question.

A number and its opposite have the same absolute value. For example $|-3| = |3| = 3$. Conversely, if $|a| = |b|$ then either $a = b$ or $a = -b$.

What are the two numbers whose absolute value is 10?
Is there a number such that if we know its absolute value we know what
it is?

If a number a is non-negative, then $|a| = a$. For this reason the
absolute value of the absolute value is just the absolute value: $|(|x|)| = |x|$.
Taking absolute value twice is the same as doing it once. It is like mashing
potatoes: mashing mashed potatoes leaves them still mashed, it doesn't do
anything new.

If a is negative, then $|a| = -a$. For example, $|-7| = -(-7)$.

In each of the following equations find a number a satisfying it (there
may be more than one correct answer):

a. $|a| = 10$ and a is positive b. $|a| = 10$ and a is negative c. $|-a| = 1$
d. $|a + 1| = 0$ e. $|a| = 0$ f. $|a - 1| = 1$ and a is positive. g. $|a - 1| = 2$
and a is positive.

Give an example of two numbers a and b satisfying the following two
conditions: $|a| > |b|$ and $a < b$.

(Possible solution: $a = -10, \ b = -5$.)

In each of the following questions complete the inequality between x
and y.
Example: If $|x| > |y|$ and both x and y are positive then $x > y$.

a. If $|x| > |y|$ and both x and y are negative then x_y.
b. If $|x| > |y|$ and x is negative and y is positive then x_y.

The Absolute Value of a Difference, and the Distance Between Numbers

For every pair of numbers x and y we have $x - y = -(y - x)$, a rule we have
met a few times. Since the absolute values of two opposite numbers are the
same, we have $|x - y| = |y - x|$. For example, $8 - 5 = 3$ and $5 - 8 = -3$, and
hence $|5 - 8| = |8 - 5|$.

Add "+" and "−" signs before the absolute values to get valid
statements:

a. $8 = \_|8|$ b. $|5 - 8| = \_|5| + \_|8|$ c. $8 = \_|8|$ d. $|-(-(-8))| = \_|-8|$
e. $|5 - 8| = \_|5| + \_|8|$

(Solution of e: $|5 - 8| = -|5| + |8|$).

Look at the two numbers whose absolute value is 5, namely 5 and (-5). Their distance from 0 is 5.

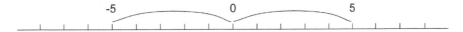

The distance of 0 from 5 is its distance from (-5), namely 5.

The absolute value of a number is its distance from 0.

In general:

The distance between two numbers x and y on the line is $|x - y|$.

For example, the distance between 5 and 8 is $|5 - 8|$, which is the same as $|8 - 5|$, which is 3.

What are the numbers whose distance from 5 is precisely 3?

(Answer: 2 and 8.)

Defining Sets of Numbers Using Absolute Values

What numbers x have absolute value less than 3, namely $|x| < 3$? Since absolute value is the distance from 0, these are the numbers whose distance from 0 is less than 3. Here they are on the line:

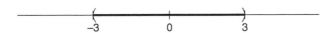

Round brackets mean that the end numbers are not included. Here, the numbers 3 and (-3) do not belong to the set. So, these are the numbers x satisfying $x > -3$ AND $x < 3$, both conditions.

The numbers x not satisfying this condition are those whose absolute value is at least 3, namely $|x| \geq 3$. These are the numbers far from 0, at least 3 units. Here is their picture on the line:

Square brackets indicate that the end numbers ARE included in the set. In this case, 3 and (-3) do belong to the set. These are the numbers x satisfying $x \leq -3$ OR $x \geq 3$, one of the two conditions.

(The notation $x \leq -3$ means that $x < -3$ or $x = -3$.)

This is a special case of a law called De Morgan's law, after the 19-th century English mathematician August De Morgan. Here is another example: if somebody promises that he will buy you ice cream AND give you a flower, not fulfilling the promise means that he or she did not buy you ice cream OR did not give you a flower. And in the other direction, if you know that the statement "Today is Tuesday OR it is raining" is false, you know that today is not Tuesday AND it is not raining.

In the picture the numbers with absolute value at most 2 are darkened. What algebraic conditions do they satisfy? Write also the conditions fulfilled by the numbers NOT fulfilling the condition.

```
-8  -7  -6  -5  .-4  -3  -2  -1   0   1   2   3   4   5   6   7   8
 |   |   |   |   |   |   |   |   |   |   |   |   |   |   |   |   |
```

Multiplication and Division of Directed Numbers

Strangely, multiplication and division are simpler for directed numbers than addition and subtraction. You just multiply or divide the absolute values, which are positive numbers so we know how to multiply or divide them, and then you take care of the sign.

Multiplying a Positive Number by a Negative Number

George owes each of his four friends 3 dollars. How much money does he owe altogether?

Of course, 4×3, namely 12 dollars. Since we are speaking of debts, this means that $4 \times (-3)$ (4 debts of 3 dollars each) is $(-(4 \times 3))$ (a debt of 4×3).
So, $4 \times (-3) = -(4 \times 3) = -12$.
In general, $a \times (-b) = -(a \times b)$.
Since multiplication is commutative, we have also $(-a) \times b = -(a \times b)$.

So multiplication of a positive number by a negative number is indeed easy, as I promised you. You just multiply the numbers, and add a minus sign.

Calculate $10 \times (-10)$ and $(-10) \times 10$. Tell an arithmetical story requiring the calculation of $10 \times (-10)$. (Example: in the Antarctics the temperature dropped 10 consecutive days, each day 10 degrees. How much did the temperature drop in these ten days together? Admittedly, not a realistic story.)

"Not Not" is "Yes", and the Multiplication of Two Negative Numbers

The rule for multiplying two negative numbers is "minus times minus is a plus". Like "not not" is "yes". So $(-3) \times (-2) = 6$.

Of course, this demands explanation. The students like most the rule "not not is yes". But here are two more explanations:

Sequences

Calculate, from left to right: $3 \times (-2) = -6$, $2 \times (-2) = -4$, $1 \times (-2) = -2$, $0 \times (-2) = 0$.

At each step the multiplier decreases by 1, and the product goes up by 2. This is clear: at each step the "debt" decreases by 2, meaning that the result grows by 2.

Going one more step, the multiplier will be 1 smaller than 0, namely (-1). The product will be 2 more than 0, namely 2. So, $(-1) \times (-2) = 2$. A minus times a minus is plus. Continuing, we get $(-2) \times (-2) = 4$, $(-3) \times (-2) = 6$, as claimed.

Using Distributivity

Look at the expression $-3 \times (-2 + 2)$. This is (-3) times 0, which is 0. Opening the parentheses by the distributive law yields then $(-3 \times -2) + (-3 \times 2) = 0$. The left parentheses, $(-3) \times (-2)$, is what we want to calculate. The right parentheses we already calculated: $(-3) \times 2 = -6$. So, $(-3 \times -2) - 6 = 0$. But $6 - 6 = 0$, so $(-3) \times (-2) = 6$.

Dividing Directed Numbers

If a debt of \$20 is divided between 10 people, each of them owes now \$2. This means that $(-20) : 10 = -2$. Namely, $(-20) : 10 = -(20 : 10)$. This is the same rule as for multiplication: you divide ignoring the sign, and then add the sign. In general:

$$(-a) : b = -(a : b)$$

Calculate: a. $(-2) : 4$ b. $(-4) : 2$ c. $\frac{-4}{2}$ d. $\frac{-x}{y}$.

And how about $20 : (-10)$? Here a story is harder to come by. There is no such things as "(-10) people" among which we can divide 20 dollars. But remember that $20 : (-10)$ is the number whose product by (-10) is 20, and it is easy to know the answer. We already know that $(-2) \times (-10) = 20$. So, $20 : (-10) = -2$.

The general rule is:

$$a : (-b) = (-a) : b = -(a : b)$$

A rule for the lazy

What is $(-1) \times (-2) \times (-3) \times (-4) \times (-5)$? The rule is: ignore the signs, and multiply the absolute values. You get $1 \times 2 \times 3 \times 4 \times 5 = 120$. Now we add the sign. There are 5 minus signs, and 5 is an odd number. So, the sign of the product is "−".

The rule: an even number of minus signs gives a "+" sign in the product. An odd number of minus signs gives the product a "−" sign.

The same is true also for division.

Directed Fractions

The first thing that should be known about directed fractions is that it does not matter where you put the minus sign. For example, $-\frac{2}{3} = \frac{-2}{3} = \frac{2}{-3}$. In the left term the sign is put in front of the fraction, in the middle it is in the numerator, and in the right it is in the denominator. These are all the same. The reason is that a fraction is really the result of division, so these are nothing but the division rules that we learnt: $-(2:3) = (-2):3 = 2:(-3)$.

Calculate $\frac{(-2) \times (-3)}{(-6)}$. Solution: without the signs we get $\frac{6}{6}$, which is 1. There are 3 minus signs, and 3 is odd, so we should add a "minus" sign. The result is (-1).

Calculate: a. $\frac{2}{3} \cdot (-\frac{3}{2})$ b. $\frac{2}{3} : (-\frac{3}{2})$ c. $\frac{2}{3} : (1 : (-\frac{3}{2}))$ d. $\frac{1-\frac{1}{3}}{\frac{1}{-2}}$ e. $-2 \times \frac{-1}{-3 \times (-4)}$.

Calculate $1\frac{2}{3} \cdot (-\frac{3}{2})$. Solution: $1\frac{2}{3} \cdot (-\frac{3}{2}) = \frac{5}{3} \cdot (-\frac{3}{2}) = -(\frac{5}{3} \cdot \frac{3}{2}) = -\frac{5}{2}$.

Addition and Subtraction of Directed Fractions

To calculate the sum or difference of directed fractions, we should just apply all the rules we have found so far. Here are a few examples.

Calculate $\frac{3}{4} - 12$.

Solution: By the rule $a - b = -(b - a)$, we have $\frac{3}{4} - 12 = -(12 - \frac{3}{4})$, which is $-11\frac{1}{4}$.

Calculate $1\frac{2}{3} - 12\frac{1}{2}$.

Solution: $1\frac{2}{3} - 12\frac{1}{2} = -(12\frac{1}{2} - 1\frac{2}{3}) = -(11\frac{1}{2} - \frac{2}{3}) = -(10 + 1\frac{1}{2} - \frac{2}{3}) = -(10 + \frac{3}{2} - \frac{2}{3}) = -(10 + \frac{9}{6} - \frac{4}{6}) = -10\frac{5}{6}$.

Calculate a. $\frac{1}{2} - 1$ b. $\frac{1}{3} - 1$ c. $\frac{1}{4} - 1$ d. $\frac{1}{2} - 2$ e. $\frac{1}{3} - 2$ f. $\frac{1}{5} - 10$.

Continue the following two sequences:

$3\frac{1}{2}, 2\frac{1}{2}, 1\frac{1}{2}, \frac{1}{2}, \text{---}, \text{---}, \text{---}, \ldots$

$3\frac{1}{3}, 2\frac{1}{3}, 1\frac{1}{3}, \frac{1}{3}, \text{---}, \text{---}, \text{---}, \text{---}, \ldots$

Solution of the second sequence: the next element is $\frac{1}{3} - 1$, that by the rules we learnt is $-(1 - \frac{1}{3})$, namely $-\frac{2}{3}$. The next element is $-1\frac{2}{3}$.

In each of the following sequences add 4 more elements:

$3\frac{1}{4}, 2\frac{1}{4}, 1\frac{1}{4}, \frac{1}{4}, \text{---}, \text{---}, \text{---}, \text{---},$

$5\frac{3}{4}, 4, 2\frac{1}{4}, \frac{1}{2}, \ldots$

$1\frac{3}{4}, 1\frac{1}{4}, \frac{3}{4}, \frac{1}{4}, \ldots$

(Hint: each element is smaller than its predecessor by $\frac{1}{2}$.)

Calculate $\frac{1}{7} - \frac{1}{6}$.

Calculate a. $\frac{3}{4} - \frac{4}{5}$ b. $\frac{4}{5} - \frac{5}{6}$ c. $\frac{5}{6} - \frac{6}{7}$. What will the next exercise in the sequence be?

(Solution: the next exercise is $\frac{6}{7} - \frac{7}{8}$, whose result is $\frac{-1}{56}$.)

Part 4

The Grammar of Algebra

An infant catches the grammar of his mother tongue from the air. Acquiring the grammar of a foreign language, on the other hand, demands hard work. Algebra is nobody's mother tongue, and learning its grammar requires effort. The grammar of algebra is called "algebraic technique", which means rules for handling algebraic expressions. "Technique" is associated with "uninspiring", which is true in this case, but it is absolutely essential for "speaking algebra".

Complex Expressions

The expression $\frac{2+1}{2-1}$ is arithmetic, since it only contains numbers. On the other hand, $\frac{x+1}{x-1}$ is algebraic, since it contains also letters. Middle school mathematics introduces two innovations: the expressions are complex, and they are algebraic. It is best to avoid tackling two difficulties at once, and hence we shall start with the first of these difficulties, namely complexity.

Divide and conquer. Try not to teach two principles at once.

Brackets — A Mathematical Box

Most arithmetic operations are defined between two numbers — they are "binary". Some, like absolute value or "minus" or taking the square root, are "unary", namely they are defined on one number. There are no elementary operations on more than two numbers. What happens when there are more than two numbers involved in the calculation? We use boxes. The result of an operation between two numbers is put in a box, and next this box is used in another operation.

"Boxes" are designated by brackets. For example, in $(2 + 3) \times 4$ we first calculate $2 + 3$. The outcome, which is 5, is put in a box, which is then multiplied by 4, to get 20. The meaning of $2 + (3 \times 4)$ is different — here we first multiply 3 by 4, and to the result, 12, we add 2 to get 14.

Brackets within Brackets

Sometimes it is necessary to put a box within a box, like these "babushkas".

To calculate the outcome, you go from inside out. Start with the innermost box, and go on to the outer ones. For example, in $((10 - 3) + 4) - (5 - 2)$ the first to be calculated is $10 - 3$. The result, 7, is added to 4, yielding 11, and then $5 - 2$, namely 3 is subtracted. We get $11 - 3 = 8$.

Types of Brackets

It is easy to match the two parts of a babushka — you go by size. In brackets this is impossible. All left brackets have the same size, and so do the right brackets. To solve this problem, we use different shapes of brackets. External brackets are replaced by square brackets, so for example $(4 + (3 + 1) \times 2) \times 3$ is written as $[4 + (3 + 1) \times 2] \times 3$. The next level of brackets is curly. For example $(2 + 3 \times (4 + 5 \times (6 + 7))) \times 8$ is written $\{2 + 3 \times [4 + 5 \times (6 + 7)]\} \times 8$. For further levels there is no accepted convention. You can invent your own. But this is rarely necessary.

Saving Brackets, First Rule: Associativity

Too many brackets are confusing, so we try to economize. The first way of doing so is a rule called "associativity". If you have three vases, one with 3 flowers, one with 4 and one with 5, then it does not matter in which order you collect all flowers: the first two vases and then the third, or the last two vases and then the first. This means that:

$$(3 + 4) + 5 = 3 + (4 + 5).$$

Indeed, on the left side we get $7 + 5$, and on the right $3 + 9$, which are both 12.

In general:

$$(a + b) + c = a + (b + c).$$

Which means that the brackets are redundant — one can simply write $a + b + c$.

Which calculation is easier: $(234 + 3) + 97$ or $234 + (3 + 97)$? Use this observation to calculate $345 + 7 + 993$.

Subtraction Inside the Brackets

If a child gets 5 dollars from his mother and 3 from his father, and he donates 2 dollars to the association for saving polar bears, it does not matter if he donates it from the sum he got from his father, or after he collected the two contributions he got. This means that: $5 + (3 - 2) = (5 + 3) - 2$, and in general $a + (b - c) = (a + b) - c$. Here, again, we can save on brackets, and simply write $a + b - c$.

In the chapter on directed numbers we learnt two more rules for saving brackets:

$$a - (b + c) = a - b - c,$$
$$a - (b - c) = a - b + c.$$

Associativity of Multiplication

The associative rule is valid also for multiplication. It does not matter in which order we perform the multiplication among three numbers:

$$(a \times b) \times c = a \times (b \times c).$$

Here is a geometric explanation why this is true, in a special case: $(2 \times 3) \times 5 = 2 \times (3 \times 5)$. Look at a box whose lengths of sides are 5 units (say, centimeters), 3 units and 2 units. Put it two positions.

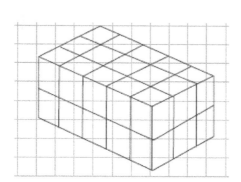
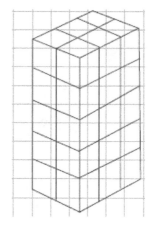

On the right, there are 5 layers, from bottom to top, each containing 2×3 small cubes. So, the box contains $(2 \times 3) \times 5$ cubes. On the left, there are 2 layers, each containing 3×5 cubes, so counting this way we get $2 \times (3 \times 5)$ cubes. Since the number of cubes is the same in both calculations, we get:

$$(2 \times 3) \times 5 = 2 \times (3 \times 5).$$

So, we can simply write $2 \times 3 \times 5$.

Conventions on the Order of Operations

Associativity saves brackets when there is only one type of operation — either addition or multiplication. For further saving, two rules have been agreed upon:

Rule (a): If only addition and subtraction are involved, or only multiplication and division, then the operations are performed from left to right.

Examples:

$$8 - 4 + 2 \text{ stands for } (8 - 4) + 2.$$

$$8 \div 4 \times 2 \text{ stands for } (8 \div 4) \times 2.$$

An important convention is:

Rule (b): Multiplication and division precede addition and subtraction.

Example: in the expression $4 + 3 \times 2$ we first perform the multiplication, $3 \times 2 = 6$, and then calculate $4 + 6$, which is 10. In $4 \div 2 + 3 \times 2$ we first calculate $4 \div 2 = 2$ and $3 \times 2 = 6$, and then add: $2 + 6 = 8$.

If we wish the addition or subtraction to come first, we use brackets. For example, in $(2 + 3) \times 7$ the addition $2 + 3$ is performed first.

These rules are arbitrary, just like the rule telling you on which side of the road to drive. But not quite. There is some logic in them. Rule (a) is the usual one, applicable in reading and in arithmetic, of going from left to right. Rule (b) says that stronger operations come first. Multiplication and division are stronger than addition and subtraction (100×100 is larger than $100 + 100$), and hence they are performed first. Why do stronger operations get precedence? This is indeed arbitrary.

Calculate: a. $1 \div (1 \div 3)$ b. $1 \div 1 \div 3$ c. $6 \div 2 + 6 \div 3$.

The Fraction Line Serves also as Brackets

Another bracket saving rule is: in a fraction the numerator and denominator get precedence. For example, in $\frac{2+5}{7-3}$ we first calculate $2+5 = 7$ and $7-3 = 4$, and then perform the division, $\frac{7}{4}$. In other words, we calculated $(2 + 5) \div (7 - 3)$, meaning that the fraction line serves also as brackets. For this reason it is usually more efficient to use the fraction line than the division sign.

> The fraction line is economical: it serves as a division sign and also as two pairs of parenthes.

A Riddle Teaching the Use of Brackets

Each of the numbers between 1 and 28 can be written using the numbers 1, 2, 3, 4, each precisely once, using the four basic arithmetical operations. For example, $10 = 1 + 2 + 3 + 4$, $11 = 1 + 2 \times 3 + 4$, $12 = 2 \times 4 + 3 + 1$, $13 = 4 \times 3 - 2 + 1$, $14 = 1 \times 3 \times 4 + 2$.

A good exercise, requiring the knowledge of the use of brackets and the order of the operations, is expressing all numbers up to 28 in this way. For example, to express 27, you need to write $3 \times (2 \times 4 + 1)$.

What is the largest number that can be represented this way?

Opening Boxes — Distributivity

Suppose you have three boxes, each containing an apple and a banana. What will you have if you open them? Of course, 3 apples and 3 bananas. In a picture:

Remember? A box is written as brackets. So, this means:

$$3 \times (a + b) = 3a + 3b.$$

It is just a coincidence that "a" stands for "apple" and "b" for "banana". They can represent anything — fruit, animals or, the most useful for us, numbers. For example, $3 \times (5 + 2) = 3 \times 5 + 3 \times 2$. And in general:

$$c \times (a + b) = ca + cb.$$

This law is called "the law of distribution". Since multiplication is commutative ($c \times d = d \times c$), the law applies also when the brackets are on the left: $(2 + 3) \times 4 = 2 \times 4 + 3 \times 4$. In general: $2x + 3x = 5x$ (2 apples plus 3 apples are 5 apples), and $ax + bx = (a + b)x$.

In fact, you used it as early as Grade 2, when you calculated multiplication. How do you calculate 34×2? You remember that $34 = 30 + 4$, and you calculate $30 \times 2 + 4 \times 2$.

The distributive law can simplify calculations. For example, just as 3 apples plus 97 apples is 100 apples, $3 \times 7 + 97 \times 7 = 100 \times 7 = 700$.

Calculate $96 \times 777 + 4 \times 777$.

The Distributive Law for Subtraction

Three containers, each containing 5 tons of bananas, were kept for too long in the port, and as a result 2 tons of bananas in each container rotted. There are two ways to calculate how many tons remained. One is gathering all bananas from all containers, and then throwing the $3 \times 2 = 6$ tons of rotten bananas; the other is throwing 2 tons from each container, and then gathering. In the first we get $3 \times 5 - 3 \times 2$ tons, and in the second $3 \times (5 - 2)$. Clearly the result is the same: $3 \times (5 - 2) = 3 \times 5 - 3 \times 2$, and in general:

$$a \times (b - c) = a \times b - a \times c.$$

Birds of a Feather Flock Together

In the zoo, they put in one cage the tigers and in another the antelopes. In algebra, we put together terms of the same type, and use the distributive law: $3xy + 2xy = (2 + 3)xy = 5xy$.

A thought principle: bring to extreme

Two cars drive from city X to city Y. Car A drives half the time with speed of 100 kmh, and half the time at speed of 50 kmh. Car B drives half the distance at speed 100 kmh and half the distance at speed 50 kmh. Which car gets faster to Y?

A basic thought principle is of help here: bring to an extreme. Think of cars that go at speeds 100 kmh and 1 kmh. Clearly, a car that drives half the way at speed of 1kmh will get to its destination VERY slowly.

Indeed, car A will arrive faster. The reason is it travels less than half the way at 50 kmh, since at speed 50 kmh it goes a smaller distance than at speed 100 kmh. So, it goes a larger distance than B at high speed.

Of course, there is nothing special in the numbers 100 and 50. The same is true for any two speeds, u and v. Writing the expressions for the times of cars A and B for these speeds, the fact that A's travel time is smaller than B's travel time is expressed by an inequality called "the harmonic average — arithmetic average inequality":

$$\frac{2}{\frac{1}{u} + \frac{1}{v}} \leq \frac{u + v}{2}.$$

Collecting "similar" terms makes life easier. Here are some examples of collecting similar terms involving fractions.

$$\frac{x}{2} + \frac{x}{2} = \frac{1}{2}x + \frac{1}{2}x = \left(\frac{1}{2} + \frac{1}{2}\right)x = x$$

$$\frac{a}{2} - \frac{a}{3} = \frac{1}{2} \times a - \frac{1}{3} \times a = \left(\frac{1}{2} - \frac{1}{3}\right) \times a = \frac{1}{6} \times a = \frac{a}{6}$$

Opening Up Two Pairs of Brackets

In the expression $(1 + 2) \times (3 + 4)$ we can open both pairs of brackets. The rule is: multiply every summand in one pair by every summand in the second, and then sum up: $(1 + 2) \times (3 + 4) = 1 \times 3 + 1 \times 4 + 2 \times 3 + 2 \times 4$. And in general:

$$(a + b)(c + d) = ac + ad + bc + bd$$

To prove this, open the brackets one by one. First give $c + d$ a new name, say x, and then on the left hand side we have $(a + b)x$, which is $ax + bx = a(c + d) + b(c + d)$. Using distribution again, we get $a(c + d) = ac + bd$ and $b(c + d) = bc + bd$, and summing up we get $ac + ad + bc + bd$, as promised.

This is precisely what you do in "long multiplication". For example, to calculate 37×45 you write: $(30 + 7) \times (40 + 5) = 30 \times 40 + 30 \times 5 + 7 \times 40 + 7 \times 5$.

And here are three algebraic examples, the first of which you may recognize:

a. $(x + 1)(x - 1) = x^2 - 1x + 1x - 1 \times 1 = x^2 - 1$
b. $(x + 3)(x - 3) = x^2 - 3x + 3x - 3 \times 3 = x^2 - 9$
c. $(x + 2)(x + 3) = x \cdot x + x \cdot 3 + 2 \cdot x + 2 \cdot 3 = x^2 + 3x + 2x + 2 \cdot 3 = x^2 + 5x + 6$

Open brackets in: a. $(x + 10)(x - 10)$ b. $(1 + x + x^2)(1 - x)$.

Short Multiplication

Two identities that follow from distribution will accompany us throughout the study of algebra. They are called "short multiplication".

The first is:

$$(a + b) \times (a - b) = a^2 - b^2.$$

For example if $a = b$ then both sides are 0 (why?) and the same is true if $a = -b$.

To understand a rule, test it in simple cases.

Check the identity when $b = 0$. What is the value of both sides in this case?

What happens when $b = 1$? Do you recognize it? (Look up the chapter "what is algebra".)

The proof of the rule is done by opening the brackets: $(a + b) \times (a - b) = a^2 - ab + ab - b^2$. The two middle terms cancel out, and we are left with $a^2 - b^2$.

A Shepherd's Story and a Geometric Proof

A shepherd is given a rope of length 40 meters, and he wishes to encompass with it a rectangular area for his sheep. He can form a square of side 10 meters, whose area is $10 \times 10 = 100$ square meters. Can he do better, that is, form a rectangle of larger area?

The answer is "no". The square is the best. In general, if the shepherd is given a rope of length $4a$ the best he can do is form a square of side a, whose area is a^2. The reason is that if he adds b to the width of the square, he must subtract for compensation b from its height, and then he has a rectangle of sides $a + b$ and $a - b$, whose area is $(a + b) \times (a - b)$, which as we know is $a^2 - b^2$, and since b^2 is non-negative this is at most a^2. In fact, it is a^2 only if $b = 0$, meaning that the enclosure is square.

Dido's Story

What is the best the shepherd can do, if he is not limiting himself to rectangular shaped enclosure? You may have guessed — the most symmetric shape, a circle.

This problem became known because of a famous story about Dido, the first queen of Carthage. Dido fled from Tyre, being persecuted by her brother, and reaching Carthage the miserly locals agreed to give her, for her money, an area that she "could encircle with an ox's skin". What she did was cut the skin into thin stripes, tied them together to obtain a long rope, and Carthage being on the seashore she used the sea as one side of the area. For the other side, she formed half a circle from the rope — the best solution.

> (*) What should the shepherd do with his 40 meters rope if he wants to form a rectangular shaped pen, and can use the seashore as one side of his pen?

(Hint — in the general case it is half a circle, so here it is half a ...)

Going the Other Direction — Factoring a Difference of Squares

"Factoring" an expression means writing it as a product of terms. The difference of squares formula is a neat way of factoring. For example, $4a^2 - 1$ is $(2a)^2 - 1^2$, which by the formula is equal to $(2a + 1)(2a - 1)$ — a product of two terms.

Another example: $\frac{y^2}{9} - 4 = (\frac{y}{3})^2 - 2^2 = (\frac{y}{3} + 2)(\frac{y}{3} - 2)$.

Write each of the following expressions as a product of two terms:

a. $(a+2)^2 - 16$ b. $x^2 - 25$ c. $\frac{x^2}{4} - 1$ d. $1 - \frac{x^2}{4}$ e. $(x+2)^2 - (x-2)^2$.

The Square of a Sum

The second abbreviated multiplication formula concerns the square of a sum. It is:

$$(a+b)^2 = a^2 + 2ab + b^2.$$

In words: the square of the sum of two numbers is the sum of their squares, plus 2 times their product.

Proof:

$$(a+b)^2 = (a+b) \times (a+b) = a \times a + a \times b + b \times a + b \times b = a^2 + 2ab + b^2$$

And here is a geometric proof:

Illustration: the square of a sum.

The square, whose area is $(a+b)^2$, divides into two squares of respective areas a^2 and b^2, and two rectangles, of area ab each.

Remember the rule "try simple cases"? Here are a few simple cases to be checked: if $a = b$ then the left hand side is $(a+a)^2 = (2a)^2 = (2a)(2a) = 4a^2$, while the right hand side is $a^2 + 2aa + a^2 = 4a^2$.

What is the result in each side in the case $a = -b$?

Here is another interesting case: $b = 1$. The formula is then: $(a+1)^2 = a^2 + 2a + 1$. For example: $101^2 = 100^2 + 200 + 1 = 10201$, $1001^2 = 1002001$.

By continuing this sequence guess, and then check your guess, what is 10001^2.

Calculate $1.1^2, 1.01^2, 1.001^2$. Connect with the previous exercise.

Putting $b = 2$ in the formula yields: $(a + 2)^2 = a^2 + 2 \times 2a + 4 = a^2 + 4a + 4$. For example: $102^2 = 100^2 + 400 + 4 = 10404$.

By the square of sums formula we have: $(2x + 1)^2 = 4x^2 + 4x + 1$. Use this to calculate 21^2, 201^2, 2001^2.

Write the formula for $(x + 3)^2$ and use it to calculate $13^2, 103^2, 1003^2$. Open brackets in: a. $\left(1 + \frac{1}{x}\right)^2$ b. $\left(\frac{1}{a} + \frac{1}{b}\right)^2$ c. $\left(2 + \frac{1}{x}\right)^2$ d. $\left(1 + \frac{1}{x}\right)^2$ e. $\left(x + \frac{1}{x}\right)^2$ f. $\left(\frac{1}{a} + \frac{1}{b}\right)^2$ g. $\left(2 + \frac{1}{x}\right)^2$.

(Solutions: b: $\frac{1}{a^2} + \frac{1}{2ab} + \frac{1}{b^2}$, d. $x^2 + 2 + \frac{1}{x^2}$.)

Look at the sequence of squares: $0^2 = 0$, $1^2 = 1$, $2^2 = 4$, $3^2 = 9$, $4^2 = 16$, $5^2 = 25$... The differences between the squares are: $1 - 0 = 1$, $4 - 1 = 3$, $9 - 4 = 5$, $16 - 9 = 7$, $25 - 16 = 9$ — so the sequence of differences is 1, 3, 5, 7, 9, ... Explain this using the formula for $(n+1)^2 = n^2 + 2n + 1$. Explain it also using the formula for $(n+1)^2 - n^2$ derived from the formula for difference of squares.

The Square of a Difference

There is also a formula for the square of a difference:

$$(c - d)^2 = c^2 - 2cd + d^2.$$

Similar to the formula for the square of a sum, only with a "minus" sign before the term $2cd$. One can prove it by opening brackets, or by using the fact that $c - d = c + (-d)$, implying, by the formula for the square of a sum:

$$(c - d)^2 = (c + (-d))^2 = c^2 + 2c \times (-d) + d^2 = c^2 - 2cd + d^2.$$

Use this formula to calculate a. 9^2 b. 99^2 c. 999^2 d. 0.9^2 e. 0.99^2 f. 0.999^2.

(Solution for b: $99^2 = (100 - 1)^2 = 100^2 - 200 + 1 = 9801$; Solution for f: $1 - 2 \times 0.001 + 0.000001 = 0.998001$.)

Identifying Squares of Sums

It is sometimes useful to identify an expression as a squares of a sum or a difference. So, let us start to practice now.

Write $1 + 2x + x^2$ as a square of a sum.

(Solution: $(1 + x)^2$.)

Write $1 - 2x + x^2$ as a square of a difference.

Write the following expressions as squares of sums or differences:

a. $x^2 + 4x + 4$ b. $x^2 - 4x + 4$ c. $4x^2 + 4x + 1$ d. $4x^2 - 4x + 1$ e. $100c^2 + 60c + 9$

(Solution for e: $(10c + 3)^2$.)

An Application to Geometry

The most applicable theorem of geometry is Pythagoras' theorem. It concerns right angle triangles, and says the following: the area of a square constructed on the hypotenuse is equal to the sum of the areas of the squares constructed on the two sides of the triangle.

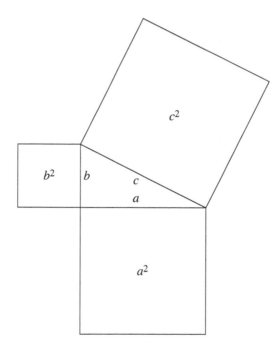

The sum of the areas of the two squares on the sides equals the area of the triangle on the hypotenuse.

Here we shall prove the theorem, using the square of sums formula. Name the lengths of the sides of the triangle a, b, c, where c is the length of the hypotenuse. The area of a square with side a (say, a centimeters) is a^2 (square centimeters). Hence the theorem says:

$$c^2 = a^2 + b^2.$$

To prove this equality, draw the following picture, where the original triangle is equal to each of the four triangles around the inside square:

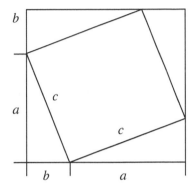

As can be seen from the drawing, the length of the side of the outside square is $a + b$, and its area is therefore $(a + b)^2$, which by the formula for the square of a sum is $a^2 + b^2 + 2ab$. On the other hand, the area of the outside square is the sum of the areas of the inside square and the four triangles. The area of a right angle triangle with sides a and b is $\frac{1}{2}ab$ (look at the picture below to convince yourself of this).

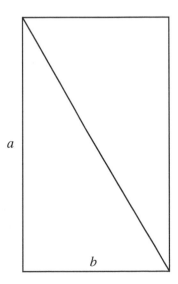

The area of the triangle is half the area of the rectangle, that is $\frac{1}{2}ab$.

Hence the area of the four triangles is $4 \times \frac{1}{2}ab = 2ab$. The length of the side of the inside square is c and hence its area is c^2. So, the sum of areas of the inside square and the four triangles is: $c^2 + 2ab$. We got two expressions for the area of the outside square, and they should be equal, so:

$$a^2 + b^2 + 2ab = c^2 + 2ab.$$

Subtracting $2ab$ from both sides we get $c^2 = a^2 + b^2$.

Calculating Distances

Why is Pythagoras' theorem so important? Because it enables calculation of distances. Here is an example.

> A man leaves his home, drives 40 km east and then 30 km north. How far, in air distance, is he from his home?

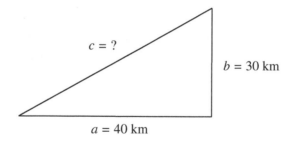

Calculating distances by the distances traveled east and north: knowing a and b yields c.

By Pythagoras, theorem, $c^2 = a^2 + b^2$, which is $40^2 + 30^2 = 900 + 1600 = 2500$.

There is only one positive number whose square is 2500, namely 50, so $c = 50$, which is the distance from point of departure.

What is the general rule? If you go x km east, and y km north, you are $\sqrt{x^2 + y^2}$ km away (air distance) from your point of departure.

Why is this so useful? Because often we know the distances east-west and south-north. For example, in Manhattan Island, the roads are woven in these directions. You know easily how far east and how far north you have gone. Distance measured by how far one goes in these two directions is in fact named in mathematics "Manhattan distance".

A Bit Beyond: Newton's Binomial Formula

Powers of 11

Write the first powers of 11, starting at 11^0. You probably remember — $11^0 = 1$ and of course $11^1 = 11$. Afterwards come: $11^2 = 121$, $11^3 = 1331$, $11^4 = 14641$. Now write them one below the other, as follows:

$$
\begin{array}{ccccccccc}
 & & & & 1 & & & & \\
 & & & 1 & & 1 & & & \\
 & & 1 & & 2 & & 1 & & \\
 & 1 & & 3 & & 3 & & 1 & \\
1 & & 4 & & 6 & & 4 & & 1 \\
\end{array}
$$

What is the rule? Every number is the sum of the two numbers above it, one to its left and one to its right. For example, $6 = 3 + 3$ and $4 = 3 + 1$. This is no coincidence. A power of 11 is obtained by multiplying the preceding power by 11, which is $10 + 1$. Multiplying by 1 contributes the left digit above, and multiplying by 10 contributes the right digit. For example,

```
        1 3 3 1
            1 1
        ----------
        1 3 3 1
      1 3 3 1
        ----------
      1 4 6 4 1
```

The rightmost 4 is obtained as the sum of 3 (from the multiplication by 1) and 1 (from the multiplication by 10), in 1331.

These are the first lines of a famous triangle, named after the French mathematician Blaise Pascal (1600–1639), but was discovered long before by the Chinese (they name it after the mathematician Yang Hui). The Pascal triangle continues by the same rule. For example, the next row is 1 5 10 10 5 1. The powers of 11 start disobeying the rule at this point, because "10" is not a digit, but a number.

We constructed the triangle from powers of 11, and indeed it serves for powers. It appears in a formula known as "Newton's binomial theorem".

The late mathematician Arnold once said that most great mathematical theorems are not attributed correctly to their inventors. Some add that even this dictum is not attributed correctly... The binomial is a case in point. Newton had important results concerning the binomial formula, notably concerning non integral powers, but he did not invent it.

The first cases of the binomial formula are:

$$(a+b)^0 = 1$$

$$(a+b)^1 = 1a + 1b$$

$$(a+b)^2 = 1 \cdot a^2 + 2ab + 1 \cdot b^2.$$

As you can see, the coefficients are taken from the Pascal triangle. Putting $a = 10$, $b = 1$ we get $11^2 = (10+1)^2 = 10^2 + 2 \cdot 10 \cdot 1 + 1^2 = 100 + 20 + 1 = 121$ — bringing us back to the powers of 11.

The next case of the binomial formula concerns $(a+b)^3$. Combine the definition of "third power" with the formula for the square of the sum to get:

$$(a+b)^3 = (a+b) \cdot (a+b)^2 = (a+b) \cdot (1 \cdot a^2 + 2ab + 1 \cdot b^2).$$

Multiplying out we get:

$$(a+b)^3 = 1 \cdot a^3 + 3a^2b + 3a \cdot b^2 + 1 \cdot b^3.$$

The coefficients, 1 3 3 1, are the next line in the Pascal triangle. And now we can guess the next case of the binomial formula, obtained from the fifth line of the triangle:

$$(a+b)^4 = 1 \cdot a^4 + 4a^3b + 6a^2b^2 + 4a \cdot b^3 + 1 \cdot b^4.$$

Write Newton's formula for $(a+b)^5$.

Part 5

Powers and Logarithms

Juggling and the Teaching of Reading

Many people know how to juggle two balls. Three is pretty hard, and four is for the advanced. Each additional ball makes the juggling many times harder. The world record, by the way, is 11 balls.

A beginner should not start from four balls, not even from two. The right number is one. The basic teaching rule "divide and conquer" reads, in this case: if there are two balls, keep one of them on the side. Start from one. If there are two ingredients to an idea, fix one of them and play with the other.

Nowhere is this more prominent than in the teaching of reading. Reading has two components: consonants and vowels. Each of them should be learnt separately. The common way to do it is to fix the vowel, and play with the consonants. Start with the vowel "a", and learn it with all possible consonants — "dad", "bad", "had", "the cat sat on the mat". This way the students learn the principle of association of a sound with a letter, without having to learn the idea of the vowel, since they are using only one vowel. Having digested this, they can move on to the next vowel — "net", "bet", "set", and "red", "bed", "shed".

The strange story of "whole words"

What is the longest way between two points?   A shortcut.

For hundreds of years the teaching of reading was going unperturbed. And then, around 1920, a novel approach took over: learning whole words. Like most educational revolutions, it started in the United States. It is possible to trace its origin to the day. It happened in 1846. A country teacher called John Russel Webb was sitting on the porch of his house, and next to him sat a little girl. In the near meadow stood a cow, grazing. Webb wrote "cow" on a sheet of paper, showed it to the girl, and pronounced the word. The girl approached her mother who was sitting nearby, and told her: "I know what is written here — 'cow'". Webb's enthusiasm knew no bounds — children can learn whole words, without going through the tedious stage of learning letters! He started using this method in his class,

(Continued)

(Continued)

and wrote a textbook about the new method. This caught up gradually, until in the 1920's it got the sanctimony of academic psychologists, who declared that first graders can learn reading by seeing whole words, just as babies learn how to speak from hearing whole words. Very quickly the method prevailed in the entire United States.

But, as shortcuts often do, this led to disaster. The method suffices for learning a few words, but when it comes to the entire language, it necessitates learning thousands of words instead of one principle. The failures led to the establishment of a whole industry of "corrective reading". In the 1990's the method took a turn to an extremist approach called "whole language", in which reading became secondary to "self expression", and this was in fact fortunate: it led to a rebellion of parents, that eventually led to the banning of the method, and a return to the old phonetic method.

Powers

The subject of this chapter, powers, is considered difficult, but this is mainly because its teaching often breaches the rule of "divide and conquer". Often juggling is attempted of two balls at once. Power is an operation between two numbers — base and exponent. Good teaching should fix one of them, and play only with the other. Best is to fix the base, and juggle with the exponent. With this approach, powers are not hard at all.

Base 10

As is often the case, a good way to start the teaching of powers is using series.

Use the didactic power of series: children love to discover regularity in series.

Start with a list:

10 to the power of 1, written as 10^1, is 10, so $10^1 = 10$. Here 10 is the base, and 1 the exponent.

10 to the power of 2, written as 10^2, is 100, so $10^2 = 100$. Here 10 is the base, and 2 the exponent.

The sequence continues $10^3 = 1000$, $10^4 = 10000$ — can you find the rule? Of course: $10^5 = 100,000$, $10^6 = 1,000,000$. Each term is 10 times larger than its predecessor, so the sequence is:

$$10^1 = 10, \ 10^2 = 10 \times 10 = 100, \ 10^3 = 10 \times 10 \times 10 = 1000,$$

$$10^4 = 10 \times 10 \times 10 \times 10 = 10,000 \ldots$$

Now we see the rule: in 10^4 the base 10 is multiplied by itself 4 times, the number of times indicated by the exponent.

Powers of 10 are simple: in 10^3 there are 3 zeros, in 10^4 there are 4 zeros, so in 10^6 there are 6 zeros. It is 1,000,000, namely a million.

How many zeros are there in $10^7 \times 10$? And in $10^7 : 10$?

The Decimal System

We started with powers of 10 for good reason. They are used in the decimal system, that we use to represent numbers. The decimal system was an idea of the ancient Indians, one of the most important in the history of mathematics, since it enables short representation of large numbers. For example,
$4635 = 4 \times 10^3 + 3 \times 10^2 + 6 \times 10^1 + 5 \times 10^0$. Powers grow very fast, so a large number can be represented by a few digits. In this system $1000 = 1 \times 10^3 + 0 \times 10^2 + 0 \times 10^1 + 0 \times 10^0$, which is just $1 \times 10^3 = 10^3$ (the rest are zeros), so $1000 = 10^3$. This is the reason that in the decimal system the representation of powers of 10 is simple.

Googol

"Googol" (not to be confused with the search engine "Google") is 1 with 100 zeros to its right, namely 10^{100}. This is possibly the largest number having a commonly used special name, but not the largest number in existence.

How many zeros are there in 10 times googol? In 100 times googol? How do you write a googol divided by 100?

Base 2

A less familiar base, but still easily negotiable, is 2.

2 to the power of 1 is 2. We write this as: $2^1 = 2$. 2 is the base, 1 the exponent.

2 to the power of 2 is 4. $2^2 = 4$.

2 to the power of 3 is 8, $2^3 = 8$.

$2^4 = 16$.

Each result is twice its predecessor. This means that at each step we multiply by 2.

$2^1 = 2$, $2^2 = 2 \times 2$, $2^3 = 2 \times 2 \times 2$, $2^4 = 2 \times 2 \times 2 \times 2$.

In 2^k you multiply 2 by itself k times.

Write $2^4 \times 2$ in power form

(Solution: $2^4 \times 2 = 2^5$).

Write $2^4 \times 2 \times 2$ and $2^{100} \times 2$ in power form.

Powers as Abbreviated Multiplication

Multiplication is abbreviated addition: 3×4 is $4 + 4 + 4$. Likewise, powers are abbreviated multiplication. Instead of $2 \times 2 \times 2 \times 2$ we abbreviate and write 2^4. The "4" tells how many time 2 is multiplied by itself.

In a^k the number a is the base, k is the exponent. The power is a product of a by itself k times. The notation was suggested by Descartes, in the 17-th century.

The Richter Scale

The strength of earthquakes is measured by the "Richter scale". The special thing about this scale is that when it goes up by 1, the strength of the tremor goes up 10 times. A scale 3 tremor is hardly felt. Scale 4 is 10 times stronger, dishes would rattle. In scale 5 buildings sway. Scale 6 is 100 times stronger than scale 4. The strongest earthquake measured was in Chile in 1960 — 9.5, and left 2 million people homeless.

How many times weaker is a 1-tremor than a 5-tremor?

(Answer: 10,000.)

Is there a tremor of strength 0? How about strength (-1)? How many times is a (-2) tremor weaker than a 4-tremor?

Noise is measured in "decibels". Similar to the Richter scale, it grows exponentially, the addition of 10 decibels means noise magnified 10 times. 80 decibels noise is 10 times stronger than 70 decibels noise.

The Legend on the Inventor of Chess

The "power" operation is not called so in vain. It is indeed strong. 2^{10} is 1024, and 2^{20} is about a million. 2^{1000} is unimaginably large — the number of atoms in the universe is about 2^{250}, and you have to double it 750 times to get 2^{1000}.

I don't know who invented the following story, but I guess it was a mathematician who tried to convey to his students how powerful is the power operation. It is about the inventor of chess, that brought his invention to the Persian Emperor, the Shah. The Shah was so enthusiastic, that he promised the inventor any gift he wished. "I don't want much", said the inventor. "I want one grain of wheat for the first square in the board, 2 for the second, 4 for the third, 8 for the fourth square, and so on". The Shah was surprised at the modesty of the request, but complied. Servants were sent to bring the grains. The 10 servant brought a small bag. The twentieth a moderate size bag. The 30-th 1024 bags, and the 40-th more than a million bags — if indeed he succeeded. And this was just the start. Very quickly the royal barns were emptied. In fact the grains corresponding to the last square — 2^{63} in number (63, not 64, because we started at 1 grain, not 2), should have been more than the number of grains ever produced on earth.

The Total Sum

How many grains should the Shah have given in all? Namely, what is $1 + 2 + 4 + 8 + \cdots + 2^{63}$? To do that, let us do smaller examples.

> "Ask me a simpler question" — tackling a problem with large numbers? Try it first in small cases.

For example, $1 + 2 = 3$, $1 + 2 + 4 = 7$, $1 + 2 + 4 + 8 = 15$. Can you see the rule? The sum is the next power, minus 1. For example, the power after $2^2 = 4$ is 8, and $1 + 2 + 4 = 7 = 8 - 1$. So, our guess will be that $1 + 2 + 4 + 8 + \cdots + 2^{63} = 2^{64} - 1$.

For the Interested — How to Prove the Formula for the Sum

Suppose that every grain in every given square sends two letters to the grains in the square following it. For example, each of the 8 grains on the 4-th square sends two letters to the 5-th square. So, every grain on the 5-th square gets one letter. Of course, the grains on the last square do not have addressees to send letters to, so they do not.

Let n be the number of grains on the board. Since every grain got a message, apart from the single grain on the first square, the number of letters received is $n - 1$. Clearly, the number of letters received is the same as the number of letters sent. The latter is $2(n - 2^{63})$, since every grain, apart from the 2^{63} grains in the last square, sent 2 letters. So,

$$2(n - 2^{63}) = n - 1$$

Or:

$$2n - 2 \times 2^{63} = n - 1$$

Solving for n gives $n = 2^{64} - 1$.

Two Balls

We learnt how to juggle one ball — the exponent. Now, to two. What is 3^4? It is $3 \times 3 \times 3 \times 3$, which is 81. Note that this is very different from 4^3, in which 4 is multiplied by itself 3 times, to give $4 \times 4 \times 4$. How do you remember which is which? The name "power" is useful here: it is the stronger number, the one telling you how many times to multiply. 2^{20} (about a million) is much larger than 20^2, which is $20 \times 20 = 400$. Look at the notation — the "stronger" number, the power, is on top!

Squaring

The power 2 is called "square", for good reason: the area of a square is the square of its side. The area of a square whose side is of length a (in, say, centimeters) is $a \times a = a^2$ square centimeters.

Base 1, Exponent 1

$1^4 = 1 \times 1 \times 1 \times 1$, which is 1. 1 to any power is 1.

Things are simple also with 1 as exponent: $k^1 = k$ for every k, because k^1 is a product with precisely one term, k.

Zero and Negative Exponents

0^k is as boring as 1^k, it is plainly 0. $0^4 = 0 \times 0 \times 0 \times 0 = 0$.

Things start getting interesting with exponent 0. Again, using sequences is useful here. Look again at the sequence

$$2^1 = 2, \ 2^2 = 4, \ 2^3 = 8, \ 2^4 = 16\ldots$$

But now look at it from right to left. Can you guess what is the next term, the one following $2^1 = 2$? Going from right to left, in each step the exponent goes down by 1, and the result goes down 2 times. So, in the next term the exponent is 0, and the result is $2 : 2 = 1$. So, $2^0 = 1$. Here is the same, with base 3:

$$3 = 3^1, \ 9 = 3 \times 3 = 3^2, \ 27 = 3 \times 3 \times 3 = 3^3, \ 81 = 3 \times 3 \times 3 \times 3 = 3^4$$

Going from right to left, the exponent goes down by 1 at each step, and the result goes 3 times down. So, the next term is $3^0 = 1$. Summarizing:

$$a^0 = 1$$

In base 10, there is a nice way to see this: 10^3, which is 1000, has 3 zeros. In general, 10^k is 1 with k zeros to its right. So, 10^0 is 1 with zero 0's to its right, namely 1.

Zero to the Power of Zero

We learnt two rules: $0^n = 0$, and $a^0 = 1$. But in one case the two conflict: 0^0. What is it – 0 or 1?

The answer is: neither. Just as 0:0 is not defined, so 0^0 is not defined.

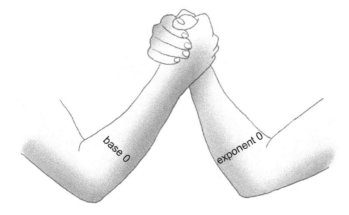

We now have at hand the sequence (read from left to right this time):

$$2^4 = 16, \ 2^3 = 8, \ 2^2 = 4, \ 2^1 = 2, \ 2^0 = 1 \ldots$$

What is the next term? It has exponent 1 less than 0, namely (-1), and the result is 2 times smaller than the result in $2^0 = 1$, namely it is 1:2. So, $2^{-1} = \frac{1}{2}$, and the next term is $2^{-2} = \frac{1}{4}$. Continuing, we get:

$$2^4 = 16, \quad 2^3 = 8, \quad 2^2 = 4, \quad 2^1 = 2, \quad 2^0 = 1, \quad 2^{-1} = \frac{1}{2},$$

$$2^{-2} = \frac{1}{4}, \quad 2^{-3} = \frac{1}{8}, \quad 2^{-4} = \frac{1}{16} \ldots$$

The symmetry catches the eye. The last term is the mirror image of the first. So,

$$2^{-n} = \frac{1}{2^n}.$$

This is true even for $n = 0$. We have $2^{-0} = \frac{1}{2^0} = 1$. Of course, this is true for every basis.

$$a^{-n} = \frac{1}{a^n}$$

What is $2^{-n} \times \frac{1}{2^n}$?
What is 10^{-6}? How do you write it as a decimal fraction?

The Rules of Powers

Product of Two Powers with the Same Base

Look at $2^3 \times 2^4$. It is $(2 \times 2 \times 2) \times (2 \times 2 \times 2 \times 2)$. It is the product of seven 2's, so $2^3 \times 2^4 = 2^{3+4}$. In general, $2^m \times 2^n = 2^{m+n}$. Of course, there is nothing special about the number 2. For every number a we have:

$$a^m \times a^n = a^{m+n}.$$

A special case is $m = -n$. We get then $a^n \times a^{-n} = a^{n-n} = a^0 = 1$. This is consistent with what we know: $a^{-n} = 1/a^n$.

Another familiar case: $10^3 \times 10^2 = 10^5$ means $1000 \times 100 = 100000$ — the rule of "addition of zeros" in the multiplication of powers of 10: 1 with 3 zeros times 1 with 2 zeros is 1 with $(3 + 2)$ zeros. Multiplying by a 100 adds two zeros to the number.

Why is it likely that the dinosaurs had cold blood?

The volume, and hence the mass, of a body, is proportionate to the 3-rd power of its length. the surface area is proportionate to the 2-nd power. So, the large dinosaurs had large weight, and not proportionately large surface area. So, if they had warm blood, it is unlikely that they could dispose of the heat they produced. This is one support for the theory that the dinosaurs had cold blood.

Quotient of Powers with the Same Base

A similar rule holds for division. For example, $\frac{2^7}{2^3}$ is $\frac{2\times2\times2\times2\times2\times2\times2}{2\times2\times2}$. Cancelling out, we get $\frac{2^7}{2^3} = 2^{7-3} = 2^4$. In fact, there is nothing new here: it is nothing but the rule $2^4 \times 2^3 = 2^7$. In general,

$$\frac{a^c}{a^b} = a^{c-b}$$

We can use this rule to prove once again that $2^0 = 1$. By our rule $\frac{2^3}{2^3} = 2^{3-3} = 2^0$. But $\frac{2^3}{2^3} = 1$.

Calculate: $\frac{2^{10}}{2^{11}}$, 0.001×10^5, $\frac{2^{10}\times3^{20}}{2^{11}\times3^{21}}$, $\frac{2^{10}\times2^{20}}{2^{11}\times2^{21}}$.

Same Exponent

What is $2^4 \times 3^4$? Here we have a product of terms with the same exponent, but different bases. It is $(2 \times 2 \times 2 \times 2) \times (3 \times 3 \times 3 \times 3)$. Rearranging, we get $(2 \times 3) \times (2 \times 3) \times (2 \times 3) \times (2 \times 3)$, which is $6 \times 6 \times 6 \times 6$, namely 6^4. So, we have $2^4 \times 3^4 = (2 \times 3)^4$.

In general,

$$a^k \times b^k = (a \times b)^k.$$

What is $2^6 \times 5^6$? And $4^3 \times 25^3$?

The same rule applies to quotients.
$\frac{6^4}{3^4} = \frac{6\times6\times6\times6}{3\times3\times3\times3} = \frac{6}{3} \times \frac{6}{3} \times \frac{6}{3} \times \frac{6}{3}$, which is $\left(\frac{6}{3}\right)^4$, namely 2^4. In general,

$$\frac{a^k}{b^k} = \left(\frac{a}{b}\right)^k.$$

What are $\frac{6^4}{2^3\times3^2}$, $\frac{10^6}{5^5\times4^3}$?
Prove the quotient rule from the rule on products.

Power of a Power

What is $(2^3)^4$? Let us first describe this expression in words: you first take 2 to the power of 3, and the result is taken to the power 4.

$$(2^3)^4 = 2^3 \times 2^3 \times 2^3 \times 2^3 = (2 \times 2 \times 2) \times (2 \times 2 \times 2) \times (2 \times 2 \times 2) \times (2 \times 2 \times 2)$$

How many 2's do we have in the product? 4×3, namely 12. So, it is $2^{4 \times 3}$.

In general, $(2^m)^n = 2^{mn}$, and more generally:

$$(a^m)^n = a^{mn}$$

How many 0's are there in 10^5? How many zeros are there in $(10^5)^2$? How many zeros are there in $(10^2)^5$?

The Order of the Operations, with Powers

Remember the rule? Powerful operations first. And the power, or exponentiation, is the strongest operation so far. So, it comes first. In $2^3 \times 2^4$ we first calculate the powers, $2^3 = 8$, $2^4 = 16$, and only then the multiplication. We get 8×16, which is 128.

The priority of the power is manifest in yet another rule: operations within the exponent first. For example, in 2^{3+4} we first calculate $3 + 4$. So, $2^{3+4} = 2^7$, which is 128.

By this rule, 10^{10^3} is $10^{(10^3)}$, that is, 10^{1000}.

How many zeros are there to the right of 1 in $10^5 \times 10^{2 \times 3 - 10}$?

Answer: the number of zeros is the exponent of 10, which is $5 + 2 \times 3 - 10$, which is 1.

When the Base is Negative

What is $(-1)^2$? The parentheses mean that you first take the minus of 1, and then square the number, so it is -1×-1, which is 1. $(-1)^3$ is $-1 \times -1 \times -1$, which is (-1). Just as "not not" is "yes", and "not not not" is "not".

$(-1)^n = 1$ if n is even, and (-1) if n is odd.

Similarly, $(-2)^3 = (-2) \times (-2) \times (-2) = -8$, and $(-2)^4 = 16$. What is the rule?

Combinations

One place in which powers appear naturally is combinations. For example:

How many words of length 2 are there, using the Latin alphabet?

The words need not be legal English words, anything goes — ab, cc — any word. There are 26 possibilities of choosing the first (left) letter, and in each of these 26 possibilities there are 26 ways to choose the second letter. So, the answer is $26 \times 26 = 26^2$. The number of 3 letters words is 26^3, and the number of words of length k is 26^k.

If we use the digits $0, 1, \ldots, 9$ instead of letters, then the number of possible "words" of length 3 is 10^3, namely 1000. But actually, we knew this before! These are precisely the 3-digit numbers, which are the numbers between 0 (which we write in this case as 000) and 999 — the numbers (including 0) below 1000, of which there are of course 1000.

What is the probability of getting 6, 6, 6 in three casts of a die?

There are $6 \times 6 \times 6 = 6^3 = 216$ possible outcomes of three casts, and only in one of them the outcome is $6, 6, 6$, so the probability is $1/216$, which is $\frac{1}{6^3}$.

In a class of 30 students, what is the probability that all share the same birthday?

Choose a student, say S. For each of the other 29 students, the probability that they have the same birthday is $1/365$ (ignoring 29 days Februaries). So, the probability that all were born on that day is $\frac{1}{365^{29}}$.

This is a very small probability. Surprisingly, if we demand only that two students have the same birthday, the probability is more than $\frac{1}{2}$!

Representing Large Numbers Using Powers

Powers can be large very quickly. This is why the decimal system is so successful — with 6 digits we can represent numbers up to a million!

A customary way of representing large numbers as a number between 1 and 9, times a power of 10. For example, 3×10^{20} or 6.4×10^{12}. The distance of earth from the sun is about 1.49×10^8 kilometers. The mass of earth is about 6×10^{24} kilograms, which is 6 followed by 24 zeros.

The speed of light in void is 300,000 km/sec. Write the speed in meters/sec in the powers notation.

(Answer: 3×10^8.)

There is similar notation for very small numbers — this time with negative powers of 10. For example, the mass of an electron is 9.1×10^{-31} kilograms, or 9.1×10^{-28} grams.

Roots

Addition has an obvious inverse: subtraction. The reverse of multiplication is division. What is the inverse of the power?

The surprising answer is that there are two inverse operations. The reason is that exponentiation (taking powers) is not "commutative". It has two ingredients, with different roles. In $5^2 = 25$ the roles of 5 and 2 are very different, unlike the operation $5 + 2$, in which they are exchangeable. So, we can ask two distinct questions:

(1) What, to the power of 2, gives 25?
(2) 5 to which power is 25?

The answer to the first question is called "root", and the answer to the second is called "logarithm".

In this chapter we are concerned with the first question. The number whose square is 25 is called the "square root" of 25, and is denoted by $\sqrt{25}$. The notation $\sqrt{}$ may have been born from the handwritten version of the letter "r", "root".

In fact, as we know well, there are two numbers whose square is 25, namely 5 and (-5). But we choose only one of them, 5, to be the square root. We have to make a choice, and this is the more natural (or more positive) one.

$$\boxed{\sqrt{a} = b \text{ means that } b^2 = a \text{ and } b \geq 0}$$

Another way of putting this is: if $b \geq 0$ then $\sqrt{b^2} = b$, and also: $(\sqrt{b})^2 = b$. What happens if $b < 0$? For example, $\sqrt{(-2)^2} = \sqrt{4} = 2$, so we got the absolute value of (-2). In general, $\sqrt{b^2} = |b|$.

Here are some eamples:

$$0^2 = 0, \sqrt{0} = 0$$
$$1^2 = 1, \sqrt{1} = 1$$
$$2^2 = 4, \sqrt{4} = 2$$
$$3^2 = 9; \sqrt{9} = 3$$
$$4^2 = 16; \sqrt{16} = 4$$

Continue this sequence up to $10^2 = 100$, $\sqrt{100} = 10$.

Roots of Negative Numbers

There is no number x such that $x^2 = -4$. At least, not among the numbers we are familiar with at this point, those called "real numbers". So, at least for now, $\sqrt{-4}$, and with it the root of any negative number, do not exist. Later on we shall generate numbers called "imaginary", among which we can find such roots.

Roots of Fractions

$(\frac{1}{2})^2 = \frac{1}{4}$ means that $\sqrt{\frac{1}{4}} = \frac{1}{2}$. So, also fractions have roots.

Square the following fractions: $\frac{1}{3}, \frac{1}{4}, \frac{1}{5}, \frac{1}{10}, \frac{2}{3}, \frac{3}{4}, \frac{3}{2}, \frac{8}{3}$. Use the results to take square roots of the following fractions: $\frac{4}{9}, \frac{9}{4}, \frac{1}{9}, \frac{1}{25}, \frac{64}{9}, \frac{1}{16}, \frac{9}{16}, \frac{1}{100}$.

Rules of the Root Operation

The root is the inverse of squaring. Hence, it obeys similar rules. Just like the square of a product is the product of squares, the root of a product is the product of roots, namely:

$$\sqrt{a \times b} = \sqrt{a} \times \sqrt{b}.$$

How do we know? The proof of the pudding is in its taste, and the proof of the root is in squaring. The equality means that the square of the right hand side is $a \times b$. And indeed:

$$(\sqrt{a} \times \sqrt{b})^2 = \sqrt{a}^2 \times \sqrt{b}^2 = a \times b.$$

Of course, the same is true also for quotients.

$$\sqrt{\frac{a}{b}} = \frac{\sqrt{a}}{\sqrt{b}}.$$

Calculate $\sqrt{\frac{9}{4}}, \sqrt{\frac{25}{4}}, \sqrt{\frac{4}{25}}, \sqrt{\frac{49}{100}}$.

Find a number whose square root is $\frac{2}{3}$.

The Fascinating Story of $\sqrt{2}$

It is rare to find drama in mathematics. The square root of 2 bred turmoil that, if we believe ancient sources, resulted in the death of a person.

The story is that of a rare creature — a mathematical cult. Pythagoras, the inventor of the famous theorem, was Greek, but like many of his compatriots did not live in Greece, but in Croton, a town in the south of

Apenine peninsula, what we nowadays call "Italy". He was the leader of a mathematical cult of some 600 people. They lived secludedly, giving all their worldly possessions to the cult. Eventually, the seclusion took its revenge: incited mob attacked and killed most of them.

Pythagoras believed in fractions. He thought that they possessed mysterious powers. This belief was the result of an ingenious discovery of his, that fractions play a major role in music. He noticed that sounds produced by hitting rods went well together if the ratio between the lengths of the rods was a simple fraction, like $\frac{2}{1}$, or $\frac{3}{2}$. Today we would have preferred to say it about the lengths of chords, rather than rods. And today we know the reason for the harmony between such notes — there is a simple ratio between their frequencies, and hence they share what is called "overtones".

As a true Greek, Pythagoras admired beauty, and he thought to have discovered such ratios also among the trajectories of planets. So he, and together with his cult, believed that every important number in the world is a fraction, or in present day terminology — "rational". One such "important" number is $\sqrt{2}$. It appears naturally in geometry — by the Pithagorean theorem it is the length of the diagonal of a square with sides of length 1. The people of the cult tried to express $\sqrt{2}$ as a fraction, and failed. Being good mathematicians, they discovered one day a shocking fact: that it is not a fraction. The (apocryphal) story goes that they swore not to disclose the secret, and when a member of the cult told the secret to the outside world, they killed him.

Why is $\sqrt{2}$ not a fraction? Assume it is, namely $\sqrt{2} = \frac{p}{q}$, for some natural numbers p, q, which means that $\left(\frac{p}{q}\right)^2 = 2$, namely $\frac{p^2}{q^2} = 2$, or

$$2q^2 = p^2. \tag{*}$$

The secret is now writing each side of (*) as the product of prime numbers, and looking at the power of 2 in each side. The power of 2 in a square number is even. For example, $24 = 2^3 \times 3$, having 2 to the power of 3. In its square $24^2 = 2^6 \times 3^2$, the power of 2 is 6, twice its power in 24.

So, on the right hand side the power of 2 is even.

On the left hand side q^2 contains an even number of 2's, and since multiplying by 2 adds one 2, the left hand side has an odd number of 2's.

It turns out that the numbers of 2 factors in the two sides are different, so the two sides cannot be equal.

If $\sqrt{2}$ is not a fraction, then what is it? It is a limit of fractions. $\sqrt{2} = 1.4142135623\ldots$ means that $\sqrt{2} = \lim 1, 1.4, 1.414, 1.4142, \ldots$

Third, Fourth Roots and so on

"Root" means usually "square root", but there are others. The inverse operation to taking third power is called "third root" and is denoted by $\sqrt[3]{}$. For example, $2^3 = 8$, so $\sqrt[3]{8} = 2$; $10^3 = 1000$, so $\sqrt[3]{1000} = 10$.

What is $\sqrt[3]{17^3}$? And $(\sqrt[3]{17})^3$? Generalize.

(Solution: $(\sqrt[3]{a})^3 = a$, $\sqrt[3]{a^3} = a$.)

Calculate $\sqrt[3]{10^6}$. What is $\sqrt[3]{a^{3n}}$?

(Solution: $\sqrt[3]{a^{3n}} = a^n$.)

Find a number b such that $\sqrt[3]{b} = \frac{3}{2}$.

Roots are Fractional Powers

We know what negative powers are, for example $10^{-2} = \frac{1}{10^2} = \frac{1}{100}$. What are fractional powers, say $9^{\frac{1}{2}}$?

To see this, take this number to the power of 2. We know that $(a^b)^c = a^{b \times c}$, so $(9^{\frac{1}{2}})^2 = 9^{\frac{1}{2} \times 2} = 9^1 = 9$. So, $9^{\frac{1}{2}} = \sqrt{9} = 3$. Taking a number to the power $\frac{1}{2}$ means taking its square root. $100^{\frac{1}{2}} = \sqrt{100} = 10$.

And how about power $\frac{1}{3}$? For example, what is $8^{\frac{1}{3}}$? Taking $8^{\frac{1}{3}}$ to the power 3 gives, by the same rule of exponentiation, $(8^{\frac{1}{3}})^3 = 8^{\frac{1}{3} \times 3} = 8^1 = 8$. So, $8^{\frac{1}{3}}$ to the power of 3 is 8, so $8^{\frac{1}{3}} = \sqrt[3]{8}$.

In general, $a^{1/n} = \sqrt[n]{a}$.

So, we now know how to take powers of the form $\frac{1}{n}$. What about general fractions, say $8^{\frac{2}{3}}$? Here it is: $8^{\frac{2}{3}} = 8^{2 \times \frac{1}{3}} = 8^{\frac{1}{3} \times 2} = (8^{\frac{1}{3}})^2$, which is $(\sqrt[3]{8})^2$, which is 2^2, namely 4.

Taking a fractional power $\frac{m}{n}$ means taking n-th root, and then raising to the power of m. In fact, it can be done in the other order — first raise to the power of m, and then take n-th root.

Calculate $16^{\frac{1}{4}}$, $81^{-\frac{1}{4}}$.

(Partial solution: $81^{-\frac{1}{4}} = \frac{1}{81^{\frac{1}{4}}} = \frac{1}{\sqrt[4]{81}} = \frac{1}{3}$.)

What is $1000^{\frac{5}{3}}$?

(Answer: $10^5 = 100,000$.)

Logarithms

Navigation

Before the invention of the GPS, navigation was extremely difficult. In the middle of the sea there are no points of reference, and sailors had to navigate by the sun and the stars. The astronomical observations were then translated into information on the whereabouts of the ship by complex calculations. To ease the calculations, the Scottish mathematician Napier conjured a new mathematical tool: logarithms. He did not invent them (you do not invent mathematics, you discover it), but he devised a new way to use them. The secret is that using them converts multiplication into addition, which is easier to perform.

Juggling One Ball

If you adhere to our rule — playing with only one ball, keeping the other on the ground, logarithms are not hard. A 7-th grader can understand them. Here is how I taught my daughter logarithms.

As in the case of powers, you need to choose a base. Let us start this time with base 2. I told her:

Logarithm of 2 is 1
Logarithm of 4 is 2
Logarithm of 8 is 3
Logarithm of 16 is 4 — can you continue?
She could: Logarithm of 32 is 5.

So, now, let us try to define "logarithm". Logarithm of 8 is 3, and the base is 2, what is the connection? Indeed, $2^3 = 8$, namely $8 = 2 \times 2 \times 2$. So, the logarithm of a number tells you how many 2's you have to multiply to get the number. In other words, to which power you have to raise 2.

2 is the base of the logarithm, in this case. Base, because we take it to a power.

Logarithm as the Inverse of Exponentiation

I now asked my daughter what is the logarithm of 2^{10}. Meaning — to what power should we take 2, to get 2^{10}? This is like the riddle "what was the color of the white horse of Napoleon". The answer is in the question — of course, 10.

Logarithm is one inverse of the power operation: the logarithm of $1024 = 2^{10}$ is 10. Exponentiating by 10 you get 1024, the logarithm takes you back to 10. The other inverse is, of course, the root. In the logarithm you ask "what is the exponent", in the root "what is the base".

The notation for the logarithm is *log*. The base (as it should) is denoted by subscript: $\log_2 4$ is logarithm on the base of 2, of the number 4. This is 2 — you need to take power 2 of the base in order to get 4.

Calculate $\log_2 64$, $\log_2 128$, $\log_2 1024$, $\log_2 2^a$.

Logarithms in Base 10

Base 10, like 2, is easy to take logarithms in.

$\log_{10} 10 = 1$, because $10^1 = 10$.
$\log_{10} 100 = 2$, because $10^2 = 100$.
$\log_{10} 1000 = 3$, because $10^3 = 1000$.

In general, the logarithm in base 10 of 1 with k zeros to its right is k.

What is $10^{\log_{10} 100}$?

General Bases

And now — two balls. We no longer fix the base. $\log_a b = x$ means that $a^x = b$. In other words, $a^{\log_a b} = b$.

What is $\log_3 81$?
What is $\log_a a^7$?

The Logarithm of a Product

We know that $\log_2 2^3 = 3$ and $\log_2 2^4 = 4$. What is $\log_2 (2^3 \times 2^4)$? We know that $2^3 \times 2^4 = 2^7$, so $\log_2 2^3 \times 2^4 = 7$, namely it is the sum of the two logarithms.

$$\log_2 ab = \log_2 a + \log_2 b$$

Of course, this rule has nothing to do with 2. It is true for every base.

$$\log_c ab = \log_c a + \log_c b$$

Here c is any base. This is the rule that Napier used, and which makes logarithms so useful. For many years, until the invention of the computer, engineers carried on their logarithm tables, or later — something called "slide rule" that produced logarithms. When they had to calculate some complicated multiplication $a \times b$, they looked up the logarithms to the base 10 of a and of b, took the sum s of the logarithms, and then used another table to find 10^s, which is $a \times b$. I am glad I was not an engineer at that time.

> $\log_{10} 1000 = 3$, and $\log_{10} 100000 = 5$. What is $\log_{10} 1000 \times 100000$? Does it fit the rule that the logarithm is the number of zeros to the right of the 1?
> What is $\log_a(a \times a^7)$?

The Logarithm of a Quotient

If the logarithm of a product is the sum of the logarithms, the logarithm of a quotient is obviously the difference.

$$\log_c \frac{a}{b} = \log_c a - \log_c b$$

This is not a new rule: it is the same as the product rule, as one can easily realize by moving sides. The same equality can be written as

$$\log_c \frac{a}{b} + \log_c b = \log_c a.$$

Since $\frac{a}{b} \times b = a$, this is nothing but the product rule.

> Find $\log_2 \frac{2\times2\times2}{2}$, $\log_2 \frac{2^2\times2^3\times2^4}{2}$, $\log_2 2 \times 2^2 \times 2^3 \times 2^4 \times 2^5 \times 2^6 \times 2^7 \times 2^8 \times 2^9 \times 2^{10}$.

Part 6

Equations: Going Back to Where You Started

What is an Equation

I like doing equations with 4–5 years old children. I put on the ground 5 stones, I ask them to close their eyes, and cover 2 stones with my hand. I then ask them how many stones am I covering. This is an equation: they see 3 stones, so if the number of covered stones is x, then $x + 2 = 5$. The number x is called an "unknown", for an obvious reason.

$$+ 3 = 8$$

This is the second use of letters in algebra. The first, you remember, was that of "variables", used to formulate general rules, about all numbers. An unknown does not denote general numbers, but a particular number that we don't know but nevertheless want to communicate some information about it. Like a nickname for a sought after criminal, whose identity we do not know.

The information can be given by an equation, like $x + 3 = 8$, or in words: after 3 students joined the class, there are 8 students in the class. How many were to begin with?

Not every equation has a solution — for example $x = x + 1$ does not, and there are equations, like $x = x$, with many solutions (in this case — all numbers). The equation $|x| = x$ has all non-negative numbers, and only them, as solutions.

How many apples have the same weight as the cat?

How many apples have the weight of this dog? Is there a dog like this?

In this chapter we shall learn how to solve equations like $2x + 3 = 7$, in which the unknown is multiplied by some constant number. Such equations are called "linear equations". What is the source of this name — we shall understand only later. The equation $2x + 3 = 3x + 7$ is also linear, and indeed moving sides (we shall soon see what this means) yields $-x + 3 = 7$. The equation $x^2 = 100$ is an example of a non-linear equation.

Write two linear equations in the unknown y, that both have one solution, which is $y = 10$.

A good way to understand equations is to invent them.

Guessing (or Perhaps More Than a Guess?)

Rule: before trying to solve an equation, guess the solution. It gives a sound feeling for the equation. Guessing the solution of $2x + 3 = 7$ may well lead to the right answer, and in any case it gives the feeling that the solution will not be a huge number.

After an attempt or two you will probably reach the solution — $x = 2$. The next step is plugging in the solution you found: $2 \times 2 + 3 = 7$. The guess was correct.

Before solving, guess.
After solving, plug in to check.

Let me give you a list of equations, from easy to hard(er). Please guess, and please do not skip — there is logic here.

A. $x = 0$. B. $x + 1 = 1$. C. $y + 1 = 2$. D. $y + 1 = 0$. E. $z + 3 = 5$.

F. $x + 10 = 100$. G. $y + 123 = 456$. H. $y + 1234 = 5678$.

All these equations share the same form: the unknown + some number = another number. How did you do them? I think that whether advertently or not, you used a rule, and found a method. Try this method on $y + 5678 = 6543$ — here you will probably need pen and paper.

Once you grasp a principle in samll numbers, apply it also to large numbers. This necessitates abstraction and precise wording.

Going Backwards, and the Inverse of Addition

Let me now tell you how you solved all the above equations: you went backwards.

In grades 1 and 2 I ask a student to go forward 3 steps. Now close your eyes, I tell him or her, and try to go back to your starting point. Of course, he or she will go 3 steps back. I ask: if you go up 3 stairs, how do you go back? Of course, go down 3 stairs. "To return you have to go backwards", we all repeat aloud.

My guess is that this is the principle you used to solve all the above equations.

The equation $x + 3 = 8$ means: if going up 3 stairs brings you to stair number 8, in which stair did you start? Go down 3 stairs, and find out. This means that if $x + 3 = 8$ then $x = 8 - 3$.

Similarly, if $y + 1234 = 5678$ then $y = 5678 - 1234 = 4444$.

The equation is $x + 3 = 1000$, and the solution $x = 1000 - 3 = 997$.

The Inverse of Subtraction is Addition

Now, to another type of equations. Again, one step at a time, please go in order, and try to guess the solutions.

A. $x - 1 = 0$, B. $x - 1 = 100$, C. $x - 17 = 0$, D. $x - 17 = 100$,

E. $x - 11 = 22$, F. $x - 17 = 1000$, G. $y - 1234 = 5678$.

Did you have a method, or just guesses? If you think you found a method, use it to solve $y - 5678 = 1234$.

I think you had a method, even if not explicit. If you are on stair x, you go down 3 stairs and get to stair number 5, to go back to x you go up 3 stairs. If $x - 3 = 5$ then $x = 5 + 3 = 8$. If $x - a = b$ then $x = a + b$. The inverse of subtraction is addition.

Solving equations is about going backwards.

The Inverse of Multiplication is Division

Solve:

A. $x \cdot 1 = 3$, B. $2x = 100$, C. $3x = 6$, D. $3x = 1$,

E. $10x = 20$, F. $10x = 200$, G. $10x = 2$, H. $1234x = 2468$.

If you found a method, use it to solve $5678y = 1234$ and $2y = 5$.

In the equation $2x = 100$ (namely $2 \times x = 100$) you multiplied x by 2 and got to 100. To go back to x you divide by 2. $x = 100 : 2 = 50$. The inverse of multiplication is division. The solution of $ax = b$ is $x = b : a = \frac{b}{a}$.

The Inverse of Division is Multiplication

Solve, by guessing:

A. $x : 1 = 3$, B. $x : 2 = 100$, C. $x : 3 = 2$, D. $\dfrac{x}{3} = 2$, E. $\dfrac{x}{3} = 1$,

F. $\dfrac{x}{10} = 20$, G. $\dfrac{x}{2} = 123$.

If you had a method, use it to solve $y : (1\frac{1}{2}) = 12$.

The secret in all these examples was that the inverse of division is multiplication. For example, in the equation $x : 4 = 5$ you divide a number by 4 and get 5. To get back to the number, you multiply by 4. $x = 5 \times 4 = 20$. Or take the equation $\frac{x}{10} = 20$, namely $x : 10 = 20$. To go back from 20 to x you have to multiply by 10, so, $x = 20 \times 10 = 200$.

The four operations are arranged in two pairs of inverse operations. In class we call them "to go north and south" (these are, say, addition and subtraction) and "west and east" (for multiplication and division). We practice this metaphor on the floor of the class.

Negative Coefficients

In the equation $-x = 12$ the unknown appears with negative coefficient. This is not really new: the minus sign is multiplication by (-1). To get back to x, you divide by (-1), which is the same as multiplying by (-1). You get $x = -12$.

Solve $-2x = 12$.

Solution: $x = 12 : (-2) = -6$.

Equations with Two Operations

Look at the equation $2x + 3 = 7$. We got to 7 from x in two steps. First we multiplied by 2, and then added 3. In order to return from 7 to x we have to reverse these two operations. But in which order?

Think of what you do when you dress up: you first put on socks, and then shoes. To go back, you first take off your shoes, and then the socks. Last in, first out. This is the case also in solving equations. To get back, you first subtract 3 from 7 — this returns to $2x$, and then you divide by 2, returning to x. The operations are reversed, in reverse order.

A thinking principle: extreme cases

X goes 3 kilometers south, and then 2 kilometers west. Y starts at the same point, and goes 2 kilometers west, and then 3 kilometers south. Will they reach the same point?

I am mentioning this problem, because it is good illustration for a basic thinking principle: go to extreme. Consider extreme cases. In this problem, assume that the starting point is near the north pole, and that instead of going 3 km south, X and Y go 10,000 km south — bringing them near the equator. X first goes to the equator and then moves 2 km west — this does not change much his longitude. Y, on the other hand, moves 2 km west near the north pole, which may significantly change his longitude. He then remains on the same longitude, when going south. So, they will end up in very different places.

This applies to almost any starting point, though possibly less noticeably.

Balancing Sides

The "inversion" (going backward) method works well when the unknown appears only once in the equation. But this is not always the case. I am sure that you can solve $2x = x + 1$, but not by "going backward". You cannot "return" to x. There is no single x that you can return to.

For such equations there is a more general method of solution, that is called "balancing sides". You perform the same operation on both sides. If the two sides are equal, then performing the same operation on both keeps them equal.

In this case, subtract x from both sides. On the left hand side you get $2x - x = x$, and on the right hand side just 1. So, the equation becomes $x = 1$, a self-solving equation.

This method can replace the "reversing" method. For example, we know how to solve the equation $2x = 10$ by reversing — you go back from 10 to x by dividing 10 by 2, to get $x = 5$. But the same could be described by "divide both sides by 2".

There are three principles in the balancing method.

1. Performing the same operation on both sides does not change the equation (apart from one operation — multiplying both sides by 0, which annihilates the equation.)
2. The aim is to isolate the unknown. We "peel" its surroundings until it remains by itself, and the equation is self-solving
3. The "peeling" is done by reversing operations. Addition is "peeled" by subtraction, and so on.

There is a common way of writing the simultaneous operations on the two sides. Consider

$$(x + 3) : 2 = 10.$$

To get rid of the division by 2, we multiply both sides by 2:

$$(x + 3) : 2 = 10 \quad / \times 2.$$

We get $x + 3 = 20$. Next we subtract 3 from both sides:

$$x + 3 = 20 \quad / - 3,$$

which yields $x = 20 - 3 = 17$.

> Solve, and write the "peeling" steps: $5(x + 3) = 60$, $x : 2 + 4 = 8$.
> A laborer earns n dollars a day, and his assistant earns 50 dollars less a day. In 5 days they earned together 1250 dollars. What is n?

(Solution: $5 \times [n + (n - 50)] = 1250$, dividing both sides by 5 gives $2n - 50 = 250$, adding 50 to both sides and dividing by 2 gives $n = 150$.)

Moving from Side to Side

There is a short way of describing the same action on both sides. It is called "moving sides".

Example: solve $2x + 5 = 15$. The first step is subtracting 5 from both sides. We get $2x = 15 - 5$. But look what happened to the 5: it moved from left to right, and reversed its sign. Instead of "+" it appears with "−".

This is an example of "moving sides". A term added in one side is moved to the other side, where it is subtracted. Similarly, a subtracted term will move to the other side, this time added. Similarly, a term multiplied in one side will move to the other side as dividing, and vice versa.

Let us continue the same example. We got $2x = 15 - 5 = 10$. We now divide both sides by 2. This converts the equation to $x = 10 : 2 = 5$. The 2, multiplied on the left side, moves to the right hand side, where it is a divisor.

Another Example

Solve: $(x + 3) : 2 - 5 = 2x + 10$.

$$(x + 3) : 2 - 5 = 2x + 10 \ / \ + 5$$
$$(x + 3) : 2 = 2x + 15 \ / \ \times 2$$
$$x + 3 = 2 \times (2x + 15) = 4x + 30 \ / \ - 3$$
$$x = 4x + 27 \ / \ - 4x$$
$$-3x = 27 \ / \ : (-3)$$
$$x = -9.$$

The last line is the solution.

Inequalities

Let us recall two notations. "$x < 3$" means that x is smaller than 3 — strictly so, namely equality is not allowed. If we wish to allow also equality, we write "$x \leq 3$". The latter is called "weak inequality".

Find two distinct values of x satisfying $5 < x < 7$.
Find a number a satisfying $a \geq 3$, but not $a > 3$.

Information on the unknown can be given using inequalities, just as equalities.

Mary has 80 dollars. How much do we need to give her so that she has more than 100 dollars?

Answer — $80 + x > 100$, and subtracting 80 from both sides we get $x > 20$.

This is the solution: the solution of an inequality is an equivalent inequality, in which the variable is isolated in one side, and the other side is a number.

The same technique that applies to equalities works also in inequalities, with one caveat: multiplying or dividing by a negative number reverses the inequality. For example, $3 > 2$ implies that $(-3) < (-2)$ — remember? The negative numbers are mirror land. So, the solution of $5 - x < 2$ is obtained in two steps:

$$(-x) < 2 - 5 = (-3), / \times (-1)$$

$$x > 3$$

Of course, we could get the same by moving x to the right hand side:

$$5 - x < 2 \ / + x$$

$$5 < 2 + x \ / - 2$$

$$3 < x.$$

Part 7

Similarity and Proportion

Lilliput

"Gulliver's travels" of the Irish Jonathan Swift (1750–1824) was meant to be a book of pungent social criticism. For example, to show his contemporaries the stupidity of wars, Swift writes about a war between two kingdoms, over the question of at which side should a boiled egg be opened. But the book became popular as an adventure story, mainly the part about Lilliput, the country of the very small people.

Lilliput looks just like our own world, with the difference that everything is precisely 12 times smaller. People are 12 times lower, and 12 times thinner. Their furniture is 12 times smaller than ours. Their elephants, hens and cows are 12 times smaller than those in our world. Gulliver will have a hard time spotting a Lilliputian mosquito.

Why did Swift choose the ratio 1:12?

Possibly because 1:12 is popular ratio for builders of models, like doll houses. And the reason for this popularity is that calculations are easy in the inches-feet-yards system. A foot contains 12 inches.

The mathematical terminology is that the Lilliputian world is proportionate to our own world. Two systems are proportionate if one is obtained from the other by multiplying by a fixed number. For example, the sequence 1 7 18 10 is proportionate to the sequence 10 70 180 100, since the latter is obtained from the first by multiplying by 10. In the opposite direction, the first is obtained from the second upon multiplication by $\frac{1}{10}$.

Proportion rules our life. The price of the tomatoes we buy is proportionate to their weight. The amount of food we have to prepare for dinner is more or less proportionate to the number of people coming to dine.

Formulas

It is easy to express proportionality by a formula. For example, if every kilogram of tomatoes costs 3 dollars, the price of x kilograms is $3x$ dollars. The fact that there is direct proportion between the number of cars and the number of their wheels is expressed by the fact that if the number of cars is x and the number of wheels is denoted by w, then $w = 4x$. From these examples we see: a quantity y is proportional to a quantity x if there is a fixed number a such that $y = ax$. Let me stress again: this is meaningful only if there are several values of y and several values of x.

"Direct Proportion" means that system B is bigger (or smaller) than system A by a constant, non-zero, ratio a. If the quantities in A are denoted by x and the quantities in B by y, the direct proportion is expressed by the formula $y = ax$

Similarity and its Two Meanings

Two proportionate systems are said to be "similar". Indeed, Lilliput is similar to our world, in both the usual sense and the mathematical sense. Gulliver sees it as if he looked at our world through an inverse telescope, or from afar.

There are two ways to express similarity between two systems. One is that one system is obtained from the other by multiplication by a constant. The other is that the ratios between the parts in the two systems are the same. The latter is the way we use to recognize objects. Our eyes use similarity to do that. The image of a table on our retina is two times smaller when the table is 2 times further. The eye (or rather, the brain — a lot of energy is devoted to the deciphering of signals from the eye) recognizes the two images as coming from the same object by the fact that the ratios between its parts are the same.

The fact that everything in Lilliput is smaller by the same factor than in our world means that the ratios in Lilliput are the same as in our world. For example, if the height of a grown up man in our world is on the average 180 cm, and the height of an average table is 90 cm, namely half the height of a man, in Lilliput the height of a man is $180 : 12 = 15$ cm, and the height of a table is $90 : 12 = 7\frac{1}{2}$, again half the height of a man. In our world the head of a man is about one eighth of the length of his body, and in Lilliput this is the same — the ratio between the head and the body is 1:8.

Similarity of Shapes

Two shapes are similar if one is a magnification of the other by some factor. For example, here are two similar triangles:

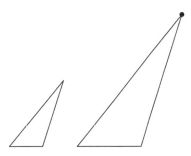

The right triangle is obtained from the left one by magnification by 2. The two triangles are similar.

A "polygon" is a closed shape formed from straight segments — for example a triangle is a polygon.

In two similar polygons not only the ratios between the sides are the same. This is true for all parts, like diagonals. And importantly — the angles are the same. Magnifying does not change angles.

Look for example at the following two similar quadrangles. They share the same angles.

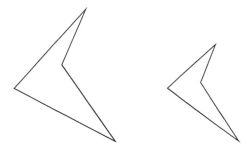

The two similar quadrangles have the same angles.

Having the same angles does not guarantee similarity. For example, a non-square rectangle has the same angles as a square — all are right angles, but they are not similar. In triangles, though, equality of corresponding angles suffices. For example, all triangles having three angles of 60° are similar. They are all equilateral.

In the following drawings there are pairs of similar shapes with some lengths of sides given. Find the missing lengths.

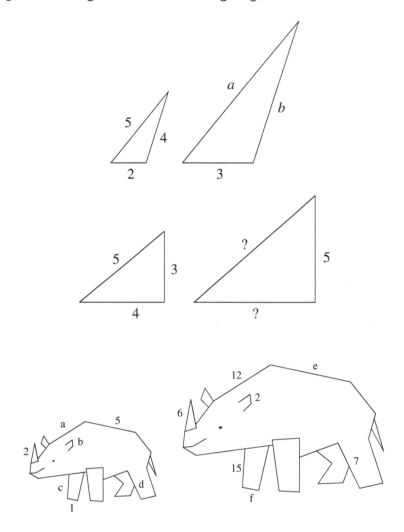

Knowing some lengths in similar shapes you can find others.

What happens to the perimeter of a square when you increase its sides 10 times? Is there direct proportion between the side of a square and its perimeter?

(Answer: yes.)

Is there direct ratio between the side of a square and its area?

(No. When you enlarge a square by a factor of 2, its area grows 4 times.)

Perfect Polygons

A polygon is called "perfect" if all its sides are equal, and all its angles are equal. For example, an equilateral triangle is perfect.

Any two perfect polygons with the same number of vertices are similar. For example, the two following perfect pentagons are similar — one is just a blow up of the other:

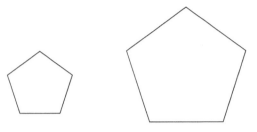

Drawing of pentagons.

In space, the analogues of perfect polygons are called "Platonic solids", after Plato, who described them. In a Platonic solid all faces are congruent (meaning, of the same shape and size), and all are perfect polygons. It was Plato who made a surprising discovery — there are just 5 such bodies! Two of them are rather familiar: the cube, and the tetrahedron (a "pyramid" with 4 vertices and 4 triangular sides). Three others are less familiar: the octahedron, having 8 triangular sides; the dodecahedron, having 12 pentagonal sides; and the icosahedron, having 20 triangular sides. Plato ascribed to these bodies celestial properties, believing they have divine properties.

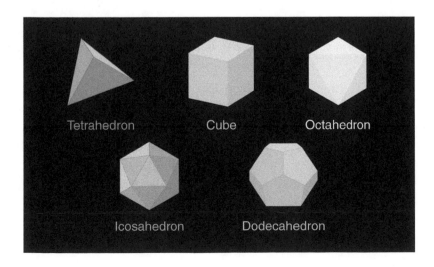

The five Platonic solids.

Similarity of Circles and the Number π

Just as any two perfect polygons of the same number of sides are similar, so are any two circles (having "infinitely many sides") similar. This means that the ratios between the different measures are the same in every circle. In particular, the ratio between the perimeter and the diameter is the same in every circle. This ratio gained a special notation — the Greek letter π. Its origin is in the word "perimeter". It is customary to estimate it at 3.14, an estimate which is less than 1% off the real value. In fact, in the 18-th century π has been proved irrational, meaning that it cannot be expressed as a fraction.

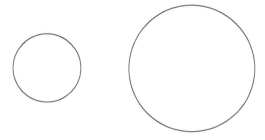

The diameter of the right circle is 2 times larger than that of the left circle. Its perimeter is in accord two times that of the left circle.

What is the diameter of circle of radius 1 meter? What is its perimeter?

Solution: Since the diameter is 2 times the radius, the perimeter is 2π times the radius.

The perimeter of a circle of radius R is $2\pi R$, which is more or less $6.28R$.

The radius of our planet is 6400 km, so its perimeter (at the equator, say) is 6.28×6400, which is 40,192 km. The "kilometer" was chosen so that the perimeter would be precisely 40,000 km, but the measurements of the French scientists of the end of the 18-th century were off by 0.5%. In fact, the French measured the perimeter along a longitude, namely from south to north, which is a bit different from the perimeter along the equator.

The Rule of Three

Given two proportionate pairs of numbers, it is enough to know three of the four numbers in order to know the fourth. Here is a simple example:

Complete the numerator of the right fraction, so as to have equality: $\frac{3}{6} = \frac{?}{10}$.

This is easy: 3 is a half of 6, so the numerator is a half of 10. The answer is 5. This is an application of the "rule of three". Another example:

A rectangle of dimensions 4 × 7 was photocopied, and in the magnified figure the width is 6 instead of 4. What is the height?

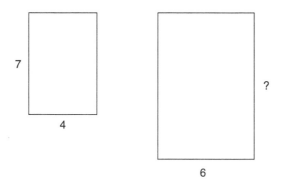

The right rectangle is similar to the left one. What is its height?

There are two approaches to the solution.

Solution A (magnifying by a fixed number): the photocopying magnified the width $\frac{6}{4}$ times, which is $\frac{3}{2}$ times. The height increased by the same ratio, so it is now $\frac{3}{2} \times 7 = \frac{21}{2} = 10\frac{1}{2}$.

Solution B (preservation of ratio): the height of the original rectangle is $\frac{7}{4}$ its width. The same should hold in the photocopied rectangle, so its height is $\frac{7}{4}$ times its width, which is 6, namely $\frac{7}{4} \times 6 = \frac{7 \times 3}{2} = \frac{21}{2}$.

A recipe for a cake calls for 2 cups of flour, one cup of oil and half a cup of water. Ben prepared the same cake, with different quantities. He used $\frac{10}{3}$ cups of flour. How many cups of oil does he need to use?

Answer (magnifying by a fixed number): $\frac{10}{3}$ is $\frac{10}{3} : 2 = \frac{5}{3}$ larger than 2, the number of cups of flour in the recipe. So, the cake is $\frac{5}{3}$ larger than in the recipe. Hence, Ben should put $\frac{5}{3}$ times as much oil, namely $\frac{5}{3}$ times 1, which is $\frac{5}{3}$ cups of oil.

Solution B (preservation of ratios): in the recipe the quantity of oil is half that of flour, so Ben should use half of $\frac{10}{3}$ cups of oil, which is $\frac{5}{3}$.

The picture on the right was obtained by magnifying the one on the right. Complete the missing lengths.

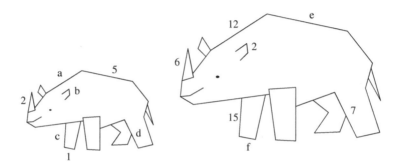

Problems of Capacity

"Bringing to a Unit"

It is rare to see the price of tomatoes in a supermarket marked as "2 dollars for $3\frac{1}{7}$ kg". Prices are marked per kilogram. This is called "bringing to a unit", in this case the unit being one kilogram. It means specifying how many units of quantity A there are for ONE unit of quantity B.

There is one place in which we are so accustomed to this approach, that we do not appreciate its wisdom – speed. Speed is indicated by "kilometer per ONE hour". Imagine how harder life would be if the speedometer would say how many kilometers the car goes every $3\frac{1}{7}$ hours!

When the proportion is not specified in this way, we should bring the proportion to a unit ourselves. For example:

For 5 shirts 3 square meters of cloth is needed. How many shirts can be produced from 12.6 square meters of cloth?

There are two ways of bringing to a unit — in one the unit is one shirt, and in the other it is one square meter. It is a matter of taste which to choose.

Solution A (one shirt): if 5 shirts use 3 square meters, one shirt needs 5 times less, namely $\frac{3}{5}$ sqm. So, the question is how many times does $\frac{3}{5}$ go into 12.6. The answer is $12.6 : \frac{3}{5} = 12.6 \times \frac{5}{3} = 12\frac{3}{5} \times \frac{5}{3} = 12 \times \frac{5}{3} + \frac{3}{5} \times \frac{5}{3} = 21$.

Solution B (one square meter): How many shirts can be made of one square meter? If from 3 sqm 5 shirts are made, 1 sqm suffices for 3 times less, namely $\frac{5}{3}$ shirts. So, from 12.6 sqm there can be manufactured $12.6 \times \frac{5}{3} = 21$ shirts.

Motion Problems: Cars Moving in Opposite Directions

Car A goes at speed 50kmh and car B goes at speed 100kmh. They leave at the same time from two cities 200 km apart, and go one towards the other. When will they meet?

Solution: going towards each other means that they approach each other at a speed which is the sum of their speeds. In this case, 150 kmh. When they meet, they have covered together 200 km, so the time it took them is $\frac{200}{150} = \frac{4}{3}$ hours.

When two cars go towards each other, their relative speed is the sum of their speeds. When they go in the same direction, their relative speed is the difference between their speeds.

Car A goes at a speed of 50 km/h, and car B at speed 80 km/h. Car B leaves 1 hour after car A, from the same point and at the same speed. How much time will elapse until they meet?

First solution: denote the time till the meeting by t. Car A will travel by this time $50t$ km, and car B $80(t-1)$ km. They will travel the same distance, $80(t - 1) = 50t$, whose solution is $t = \frac{8}{3}$. This is the time, in hours, till the meeting.

Solution B: In the first hour car A goes 50 km. From this point on, every hour the distance between the two cars decreases by $80 - 50 = 30$ km, necessitating $\frac{50}{30} = \frac{5}{3}$ hours to cover the 50 km distance. Together with the 1 hour, $t = 1 + \frac{5}{3} = \frac{8}{3}$.

Pool Problems

If you travel 60 km per hour, you travel one kilometer in $\frac{1}{60}$ of an hour; if a kilogram of tomatoes costs 5 dollars, one dollar will suffice for $\frac{1}{5}$ of a kilogram of tomatoes. In general, if the ratio between quantity A to quantity B is a then the ratio between B and A is $\frac{1}{a}$. We are not accustomed to think of speed in terms of hour per kilometer or of prices in terms of tomatoes per dollar (unless you have only 1 dollar in your pocket and you want to know what quantity of tomatoes you can buy).

There is a famous, and dreaded, type of problems. They are called "pool problems", but if well understood they are just inverse rate problems. Namely, instead of "how many units per hour", they speak about "how many hours per unit".

Here it is, in its best-known form:

Tap A fills a pool in one hour. Tap B fills it in two hours. How much time will it take the pool to be filled if both are open?

Solution: Invert. Instead of "how much time per pool", ask "how many pools per hour". The fact that the second tap fills the pool in two hours means that it fills half a pool in one hour. The first tap fills one pool an hour. So, together they fill $1 + \frac{1}{2} = \frac{3}{2}$ of a pool in one hour. So, it takes them $1/\frac{3}{2} = \frac{2}{3}$ of an hour to fill the pool together.

Do the same for two taps, one filling the pool in 4 hours and the other in 5 hours.

Here is the same problem, in travel formulation:

Car A travels the distance between Geneve and Basel in 4 hours. Car B does it in 5 hours. How much time will elapse until they meet, if they leave Basel and Geneve, respectively, at the same time, and travel towards each other?

Solution: Invert. Speak about units per hour: car A travels $\frac{1}{4}$ of the distance between Basel and Genève in one hour, and car B travels $\frac{1}{5}$ of the distance in an hour. So, together they cover $\frac{1}{4} + \frac{1}{5} = \frac{9}{20}$ of the distance every hour. So, it will take them $1/\frac{9}{20} = \frac{20}{9} = 2\frac{2}{9}$ hours to meet.

Laborer A finishes building a wall in 3 days, laborer B does it in 4 days. How much time will it take them to do the work if they join forces?

Part 8

Functions

Machines of Input and Output

A vending machine receives as input a button press, and produces an output of the appropriate drink. Imagine now another type of machine. It gets as input names of countries, and outputs ... well, you will see:

England → London

South Africa → Pretoria

Portugal → Lisbon

Given a country, the machine outputs its capital. This is an example of a "function". A function is an assignment according to some fixed rule. For every input, there is a unique output.

> A function is a machine receiving inputs, and producing appropriate outputs, according to some rule.

The input is sometimes called also the "argument" of the function.

Notation

Like everything in algebra, functions need names. As usual, the names will be Latin letters. For example, the last function can be named "C" for "capital". The common notation is then:

C(England) = London

C(US) = Washington

The input is put in parentheses, and the output is written to the right of the equality sign.

Complete: C(Iran) = ___, C( ) = Cairo, C( ) = Bogota.

Here is a function, denoted by L, receiving words as input and producing letters as output: L("word") = d, L("cat") = t, L("street") = t.

Can you guess what is L("house"), and why is the function denoted by "L"?

And why are the quotation marks there?

115

The function "N" takes words, and produces numbers. Here are some examples: N("I") = 1, N("one") = 3, N("three")=4, N("four") = 4. Can you guess the rule? What is N("unavoidable")?

> A function establishes a link between input and output. The information we have on the world is usually in the form of links, which is why functions are so important in mathematics.

Numerical Functions

The link between a country and its capital is mainly interesting for administrators, tourists or politicians. In mathematics we are mostly interested in functions linking numbers with numbers, namely numerical functions.

A numerical function is a machine "swallowing" one number, and outputting another. For example, can you guess the rule by which the following function, denoted by "f", act?

$$f(1) = 3, \quad f(5) = 7, \quad f(100) = 102$$

There are many rules by which these outputs could be produced, some simpler and some more complicated. But we shall always strive for the simplest rule possible, and in this case there is a very simple one: f adds 2 to its input. In algebraic notation:

$$f(x) = x + 2$$

Here "x" is a variable. It denotes any number.

Three Ways of Representing a Function

In the previous example we represented the function in three ways. First, we listed some of its values. Then we wrote its rule in words: "it adds 2 to its input". And finally, we gave it an algebraic formulation — $f(x) = x + 2$, certainly the most succinct one.

Express in words and in a formula the action of the function described by the following table. Also, complete the table (note that in one of empty entries you are asked for the output when the input is 1000, in the other for the input when the output is 20).

Input	1	2	3	4	_	10	100	1000
Output	4	6	8	10	20	22	202	_

It is easiest to guess by the value for 100 — obviously 202 is obtained from 100 by multiplying by 2, and then adding 2. Checking the other values shows that this rule applies there also. If we denote the function again by f, we have

$$f(x) = 2x + 2.$$

> To guess a function, it is often helpful to know its values on powers of 10 — the inputs 10, 100, 1000 and so on.

Solution to the input question: you are asked to find a value of x for which $2x + 2 = 20$. This is an equation, whose solution is $x = 9$. The input is 9.

Find a formula for $f(x)$ conforming with the following data:

$f(10) = 33$, $f(100) = 213$, $f(1000) = 2013$, $f(10000) = 20013$.

What does the "machine" of the function do?

(Answer: it multiplies the input by 2 and adds 13.)

Complete the following tables, and find formulas for the functions:

Input	1	5	10	20	200
Output	20	60	110	210	_

Input	1	2	3	4	10	100
Output	4	10	16	22	58	_

Guess what is the rule governing the action of the following function s by its following values: $s(0) = 0$, $s(1) = 1$, $s(2) = 4$, $s(3) = 9$, $s(4) = 16$.

Write an algebraic expression for s and find $s(10)$.

The function G satisfying $G(0) = 0$, $G(1) = 0$, $G(17) = 0$ and in general $G(x) = 0$ for all x is a legitimate function — although a very lazy one, and boring. A function outputting the same number regardless of the input is called a "constant function".

Functions Expressing Direct Proportion

Functions express links, and one link that we have met is proportionality.

The price of gasoline is $2.5 per liter. Let p(x) be the amount paid, in dollars, as a function of the amount of gasoline bought, in liters.

Solution: $p(x) = 2.5x$.

A function expressing proportionality is of the form $f(x) = ax$, when a is a fixed number. In the above example $a = 2.5$.

Find an expression for the function having the following table of values. Is it a proportionality function? Could you see the fact that it is not a proportionality function from just one value?

Input	0	1	2	3
Output	3	7	11	15

Solution: $f(x) = 4x + 3$. It is not a proportionality function, since a function $f(x) = ax$ satisfies $f(0) = 0$, while the value of this function at 0 is not 0.

Linear Expressions and Linear Functions

The function $f(x) = 4x + 3$ is almost a proportionality. But for the constant term 3, it would indeed be proportionality. Such a function is called "linear", for reasons to be understood later, when we get to draw functions. The number 4 is said to be the coefficient of x and 3 is the "free term", free because it stands alone, free of variables.

In general, a function is called "linear" if it is of the form $f(x) = ax + b$, where a and b are some numbers. The expression $ax + b$ is called a "linear expression".

Examples: in $-2x + 10$ the coefficient of x is (-2), and the free term is 10. In $\frac{1}{2}x - \frac{1}{4}$ the coefficient of x is $\frac{1}{2}$ and the free term is $(-\frac{1}{4})$.

Among the following expressions find the linear ones, and write the coefficients of the variable and the free terms. Remember: $\frac{x}{2} = \frac{1}{2}x$.

a. $\frac{x+1}{x}$ b. $\frac{x+2}{3}$ c. $-\frac{1}{2}x + 1$ d. $2x - x^2$ e. $\frac{1}{x}$ f. 17.

(Answer: the linear ones are b, c and f. In f the coefficient of x is 0, and the free term is 17.)

Rate of Change

How to Recognize Linearity

The function $f(x) = 2x$ is linear.

Input	1	2	3	4	5	6
Output	2	4	6	8	10	12

At each step, the function grows by 2. For example, $f(2) - f(1) = 4 - 2 = 2$, and $f(3) - f(2) = 6 - 4 = 2$. This is no coincidence. Since $f(x) = 2x$ we have $f(x + 1) = 2(x + 1) = 2x + 2 = f(x) + 2$, meaning that $f(x + 1)$ is larger by 2 than $f(x)$.

Something similar happens with the function $g(x) = 2x + 3$. Here is its table:

Input	1	2	3	4	5	6
Output	5	7	9	11	13	15

Every step the function grows by 2. Since g is obtained from f by adding 3 to everybody, the differences remain the same (if you got 2 points less than your friend in an exam, the difference will remain the same also when 3 points are added to both grades).

Linearity and Constant Rate of Change

Look at the function $g(x) = x^2$. How much does it change when x grows by 1? The answer depends on where you are. Moving 1 unit from $x = 0$ to $x = 1$ brings the function from $0^2 = 0$ to $1^2 = 1$, so the increment is 1. Moving 1 unit from $x = 1$ to $x = 2$ takes the function from $1^2 = 1$ to $2^2 = 4$, with increment 3.

Linear functions are characterized by constant rate of change. But let us first define "rate of change".

Definition: the rate of change of a linear function $f(x) = ax + b$ is the increment in the value of the function when x grows by 1, namely $f(x+1) - f(x)$.

In a linear function $f(x) = ax + b$ the increment is independent of the point x where the rate of change is measured. Why? Because $f(x+1) - f(x) = a(x+1) - ax = ax + a - ax = a$. The rate of change is simply the coefficient of x.

> Linear functions are characterized by constant rate of change. The rate of change of a linear function $f(x) = ax + b$ is a.

The rate of change of a non-linear function is called its "derivative". The rate is then not necessarily constant. Its precise definition, that you will later learn under the title of "calculus", is a bit more complicated. It takes into account moving by small jumps, not necessarily by a jump of 1.

Find the rates of change in the following functions: a. $f(x) = 3x + 4$ b. $g(x) = 3x$ c. $h(x) = 100$ d. $k(x) = -3x + 4$.
What does zero rate of change mean? What does negative rate of change mean?

Speed as a Rate of Change

A familiar example of rate of change is speed.

A car starts its journey 50 km away from the city, and goes away from the city at a speed of 80 kmh. What is its distance from the city after t hours?

Answer: in t hours it goes $80t$, so its distance from the city after t hours is $80t + 50$ km.

In this example we can see that the coefficient of the variable t is the rate of change of the way passed, namely the speed.

A General Formula for the Rate of Change

To find the speed of a car, you divide distance by time. A car traveling at constant speed 100 km in 2 hours travels $\frac{100}{2} = 50$ km in one hour. If the car travels D km in a time span T, it travels $\frac{D}{T}$ km in one hour. So, the speed of the car, which is the rate of change of the distance, is the change in distance divided by the change in time. Similarly, if adding 5 kg of tomatoes to your shopping bag added \$2 to your pay, the rate $\frac{pay}{weight}$ of change in your pay is $\frac{5}{2}$, meaning simply that the price of tomatoes is \$2.5 per kg.

A linear function f satisfies $f(10) = 100$, $f(2) = 20$. What is its rate of change?

Solution: adding 8 to x results in a change of $100 - 80 = 20$ in the function. So, adding 1 would change the function by $\frac{80}{8} = 10$ units. The rate of change is 8.

> The rate of change of a linear function is the change in the function, divided by the change in the variable.

It is known that a linear function $f(x) = ax + b$ satisfies $f(102) = 2000$, $f(100) = 1000$. What is a?

Answer: When the variable changes by 2 units, the function changes by $2a$, $f(102) - f(100) = 2a$, so $2a = 2000 - 1000 = 1000$, meaning that $a = 500$.

What does "zero rate of change" mean? Give an example of a function
with zero rate of change.

The Free Term in a Linear Expression

How does one find the free coefficient? For example, what is b in the previous
problem? Easy: put the value $a = 500$ in the equation $f(100) = a \times 100 + b$
and use the fact that $f(100) = 1000$ to obtain $500 \times 100 + b = 1000$, to
get $b = 50,000 - 1000 = 49,000$. (We could just as well have put the value
$a = 500$ in the expression for $f(102)$.)

But the free coefficient has another interpretation: it is the function at 0.
Putting $x = 0$ in $f(x) = ax + b$ we get $f(0) = b$.

Let $f(x)$ be a linear function. Find an expression for f if it is known that
$f(1) = 10$ and $f(2) = 13$.

Solution: the rate of change is $f(2) - f(1) = 13 - 10 = 3$. So, $f(x) = 3x + b$
for some b. To find b calculate $f(0)$. Since the rate of change is 3, we have
$f(0) = f(1) - 3 = 10 - 3 = 7$. So, $f(x) = 3x + 7$. We could also use the
previous way: putting $x = 1$ gives $f(1) = 3 \times 1 + b = 10$, yielding again $b = 7$.

Entrance to an amusement park costs \$5, and the use of every facility
costs an extra \$2. Write an expression for the price paid by a visitor who
used n facilities.

(Answer: $5 + 2n$.)

The grade in exam is given by the expression $100 - 10m$, where m is the
number of mistaken answers. What is the interpretation of the coefficient
10, and how many questions were there on the exam?
The height of a giraffe measured in cm, at age y (measured in years, and
assuming y is no more than 5) is $150 + 10y$. What is the interpretation
of the free term 150, and what is the interpretation of the coefficient 10?

Part 9

When Algebra Meets Geometry

One picture is worth a thousand words.

The Difference Between Road and Sea

A driver stranded on the M1 road leading from London west can specify his location for the rescuers by telling them his distance from London. One number suffices, because the road is one dimensional. In fact, this is the meaning of being one dimensional. A ship at sea, on the other hand, needs to send two coordinates, its longitude (east-west coordinate) and its latitude (north-south coordinate). The sea is two dimensional.

> There is a natural North-South point of reference — the equator. Latitude measures the distance in angles from the equator. 0 being the equator itself, 90 being the north pole, and (−90) the south pole.
>
> In the East-West direction there is no natural point of reference. In 1884, when England was still an empire, a conference was held in London to determine such a point, and the English persuaded the others that the center of the world, at least as far as the East-West directions are concerned, is London, in fact a certain part of London called Greenwich. Longitude is measured with respect to this point. So, Greenwich is at longitude 0.

A Coordinate System

Longitude and latitude are not perfect examples for the description of location by two numbers, for two reasons. One is that not every point necessitates two numbers. The north pole is the only point with latitude 90, and the south pole is the only place with latitude (−90). Worse than that, not every longitude or latitude are possible — there is no point at latitude 100.

So, let us change the story, from a ship at sea to an ant. Imagine a grain of wheat somewhere on a sheet of paper, and an ant, that wants to communicate to its friends the location of the grain. How should she do it? Chess players know the answer: use a grid. They assign letters to columns of the board, and numbers to the rows. In mathematics we only use numbers. We draw

two lines called "axes", one horizontal and one vertical, and divide each axis
by numbered marks, then the ant can tell her friend "go 6 units to the right,
and 5 units up". Or, if more mathematically versed, "go horizontally 6 units
in the positive direction, and vertically 5 units in the positive direction".

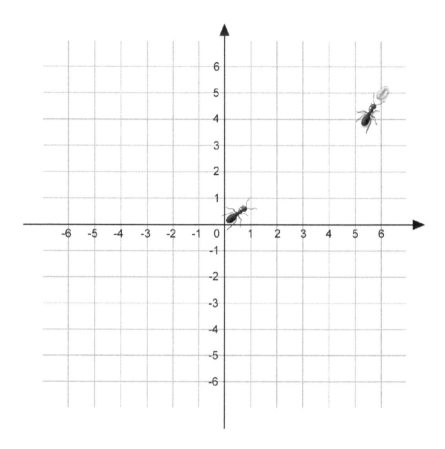

The numbers describing the location are called the "coordinates" of the
point. It is customary to call the horizontal coordinate x, and the vertical y.
The point is then denoted by (x, y). For example, the grain is at the point
$(6, 5)$. The left coordinate is the horizontal, x coordinate, the right is the
vertical, y coordinate. In accord, the horizontal axis is called the "x axis",
and the vertical the "y axis".

The point $(0, 0)$ has a special significance — it serves as a point of
reference, and it is called the "origin". Negative value of x means going
to the left of the origin, and negative value of y means going below the
origin. For example, to go to the point $(-3, -2)$ you go 3 units to the left of
the origin, and 2 units down.

Here are the drawings of some points:

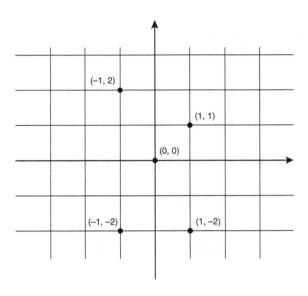

A few points and their coordinates

Draw in the plane the points $(1, 2)$, $(-1, 2)$, $(1, -2)$, $(-1, -2)$.
Write the coordinates of the points in the drawing:

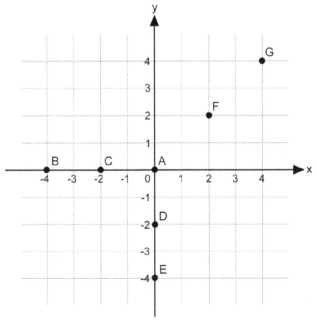

Answer: E is $(0, -4)$, G is $(4, 4)$.

I am Drawn, Therefore I am

The idea of two coordinates appeared first in the book "On the Method" of the French mathematician and philosopher René Descartes — the same book in which he wrote his famous "I think therefore I am". Descartes' Latin name was Cartesius, and in his honor the coordinate system is called "Cartesian".

René Descartes, 1596–1650

Getting from One Point to Another

To get from the origin to $(10, 10)$ you go 10 units right and 10 units up. How do you get from $(3, 6)$ to $(10, 10)$? Of course, $10 - 3 = 7$ units right, and $10 - 6 = 4$ units up.

In general, to get from point (a, b) to (c, d) you go $c - a$ to the right, and $d - b$ up. But remember: going up (-3) units means going down 3 units, and going right (-5) units means going left 5 units.

A worm starts at point $(1, 2)$, goes 3 units right, then 3 units up, (-1) units right. Where did it land?

How do you get from $(2, 3)$ to (x, y)? How do you get from (x, y) to $(2, 3)$?

Distance from the Axes

To get to the point $(6, 5)$ the ant in the story went first 6 units to the right, so 6 units away from the y axis. The distance of the point $(6, 5)$ from the y axis is therefore 6 (see drawing below). Similarly its distance from the x axis is 5.

And how about the point $(-6, 5)$? It has the same distance from the y axis, the difference being its position — it is to the left of the axis.

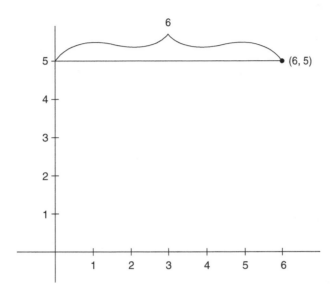

The moral is that the distance of a point (x, y) from the y axis is $|x|$, the absolute value of its x value. Similarly, the distance of a point from the x axis is $|y|$.

Points on the x axis are of the form $(x, 0)$, and indeed their distance from the x axis is 0. Points on the y axis are of the form $(0, y)$.

Find all points that are of distance 5 from the x axis and of distance 3 from the y axis.

(Hint: there are four such points.)

The Distance Between Two Points

"Distance" is a basic geometric notion, so it must have a simple algebraic expression. Indeed, it does. Let us start with the simplest — the distance of a point from the origin. Looking at the drawing below, you see: the distance of a point (x, y) from the origin is the hypotenuse of a right angle triangle, whose two sides are of lengths x and y. So, by Pythagoras' theorem, the distance d satisfies $x^2 + y^2 = d^2$, and taking roots on both sides we see that $d = \sqrt{x^2 + y^2}$.

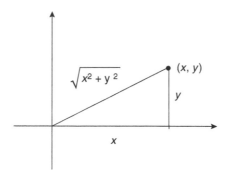

Example: the distance of the point $(4, 3)$ from the origin is $\sqrt{4^2 + 3^2} = \sqrt{16 + 9} = \sqrt{25} = 5$.

Phythagorean triples

A kind of a miracle happened in the last example: the point (4,3) has integral coordinates, and its distance from the origin is also an integer, 5. This happen rarely: for example the distance of (1,1) from the origin is $\sqrt{1^2 + 1^2} = \sqrt{2}$, which is not an integer (and as we have seen, not even a fraction, it is "irrational"). But it happens. For example, the distance of (12,5) from the origin is $\sqrt{12^2 + 5^2} = \sqrt{144 + 25} = \sqrt{169} = 13$. This means that $3, 4, 5$ are the lengths of the sides of a right angle triangle, and so are the lengths $5, 12, 13$. Such triples of integers a, b, c, satisfying $a^2 + b^2 = c^2$, are called "Pythagorean triples".

   A cheap way of getting a new Pythagorean triple is taking a known one and multiplying its elements by the same number. For example, multiplying the elements of the triple (3, 4, 5) by 10 gives $30, 40, 50$ — the distance of (40, 30) from the origin is 50. But there are infinitely many non-proportionate triples, a fact that was known already to Diophantus, in the 3-rd century BC.

What is the distance of the point $(4, -3)$ from the origin?

What is the distance between the points $(1, 2)$ and $(4, 5)$?

   To go from $(1, 2)$ to $(4, 5)$ you go $4 - 1 = 3$ units right, and $5 - 2 = 3$ units up. So, by Pythagoras theorem (see drawing below) the distance is

$\sqrt{3^2 + 3^2} = \sqrt{9 + 9} = \sqrt{18}$. There is no integer root of 18, so the answer is best left this way.

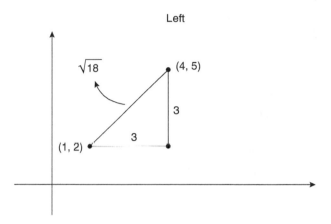

Left

In general, the distance between two points (a, b) and (c, d) is $\sqrt{(c - a)^2 + (d - b)^2}$.

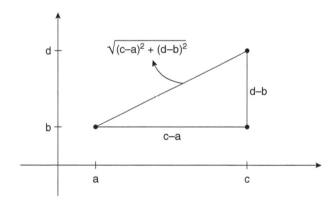

What is the distance between $(2, 5)$ and $(2, 8)$?

(Solution: by the formula, the distance is $\sqrt{(2 - 2)^2 + (8 - 5)^2} = \sqrt{3^2} = 3$.)

Of course, there is no need in this case in the formula: to go from $(2, 5)$ to $(2, 8)$ you just go up 3 units.

What is the distance between $(1, 5)$ and $(-3, 8)$?

(Careful with the signs... The answer is 5.)

Geometric Locations

Geometry Becomes Algebra

Have you ever asked yourselves what kind of commands make your computer draw the figures in your favorite game? They are neither verbal, nor given by way of a drawing. They are mathematical. In the computer, drawings are algebra. Descartes didn't have this particular aim in mind, but he did want to turn geometry to algebra.

But let us start with simple drawings, like lines or circles, rather than the complex figures in games. A line or a circle are examples of "geometric locations". A geometric location is a collection of points satisfying some simple property. For example, the collection of points satisfying the property of "being at distance 4 from the origin" is a circle of radius 4 with center at the origin. Our aim is to write the properties algebraically, instead of verbally. Let us indeed start with the circle.

Circles

Here is the circle discussed in the previous paragraph, drawn in a coordinate system. Recall that the distance of a point (x, y) from the origin is $\sqrt{x^2 + y^2}$. So, a point is on the circle if and only if $\sqrt{x^2 + y^2} = 4$, which upon squaring both sides means $x^2 + y^2 = 16$. This is the equality characterizing the points (x, y) on the circle: the geometric condition is now expressed algebraically. The equality expresses a link between the coordinates of the points on the geometric location.

> What is the equality characterizing the points on a circle of radius 3, centered at $(1, 2)$? Solution: we saw that the distance between a point (x, y) and the point $(1, 2)$ is $\sqrt{(x - 1)^2 + (y - 2)^2}$, so the equality is $\sqrt{(x - 1)^2 + (y - 2)^2} = 3$, which can also be written as $(x - 1)^2 + (y - 2)^2 = 9$.

In general, the equality characterizing the circle of radius R centered at (a, b) is

$$(x - a)^2 + (y - b)^2 = R^2.$$

Let us now go in the other direction — given an algebraic link between the coordinates, we shall try to guess the geometric location it describes.

> What is the geometric location characterized by $x^2 + 6x + y^2 - 2y = 39$?

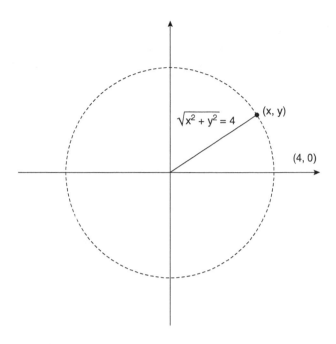

A drawing of the circle in a coordinate system

Answer: Recall that $(x + 3)^2 = x^2 + 6x + 9$, which resembles very much the left hand side of the equality. So, $x^2 + 6x = (x+3)^2 - 9$, and for a similar reason $y^2 - 2y = (y - 1)^2 - 1$. Hence the left side of the equality is:

$$x^2 + 6x + y^2 - 2y = (x + 3)^2 - 9 + (y - 1)^2 - 1$$
$$= (x + 3)^2 + (y - 1)^2 - 10,$$

and the equality is: $(x + 3)^2 + (y - 1)^2 - 10 = 39$, which is the same as $(x+3)^2 + (y-1)^2 = 49$. This is the formula of a circle centered at $(3, 1)$ and having radius 7.

Lines

Even more basic than circles are straight lines. To find how their formulas look, let us examine a few examples.

1. The equality $y = 3$ is satisfied by the points on a line parallel to the x axis, 3 units above it.
2. The points (x, y) satisfying $x = 2$ lie on a line parallel to the y axis, 2 units to its right.

3. Look at the points $(0,0)$, $(1,1)$, $(-3,-3)$, $(5,5)$. The drawing below shows that they lie on a straight line, the bisector of the angle between the axes. All these points are such that $y = x$. This is their characterizing equality.

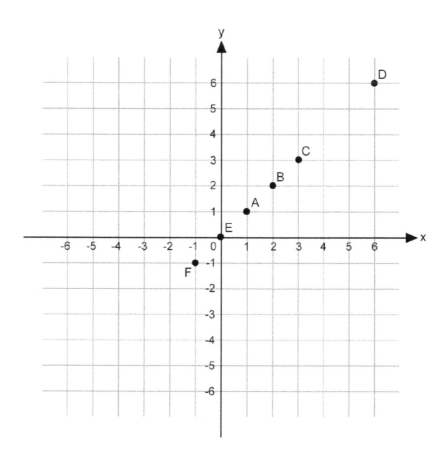

Can you describe in words the geometric location of these points? And by formula?

4. What is the property of the following points?

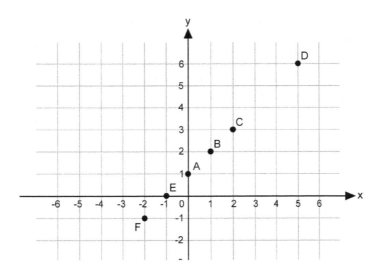

Let us first fill the gaps — we can draw a straight line through these points:

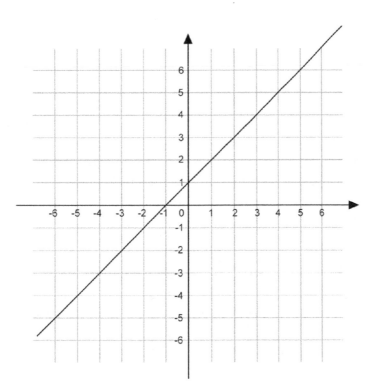

The line consists of all points (x, y) satisfying $y = x + 1$.

5. Let us next start from a formula: $y = 2x + 3$. Here are some points satisfying this condition: $(0, 3)$, $(1, 5)$, $(2, 7)$, $(3, 9)$. As you see the increment is constant — when x grows by 1, y grows by 2. Familiar? Of course, it is from the previous chapter, on linear expressions. The rate of growth is the coefficient of x.

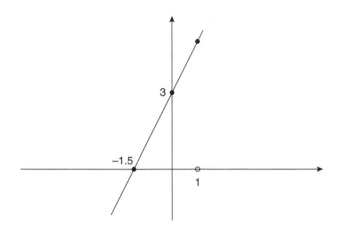

Drawing of the line $y = 2x + 3$

Lines and Linear Expressions

We now realize why expressions of the form $ax + b$ are called "linear" — because the equality $y = ax + b$ defines a straight line. But not every straight line can be described in this way. There is one type of exceptions: lines of the form $x = 3$, namely lines parallel to the y axis.

But there is a general form, in which every line can be written. We call it the "standard form". Here it is:

$$Ax + By = C.$$

Here A, B and C are any numbers, with the sole condition that not both A and B are 0. For example, putting $A = B = 1$ and $C = 5$ we get the equation $1x + 1y = 5$, meaning $x + y = 5$.

And how about the line $x = 3$? Just write it as $1x + 0y = 3$, so $A = 1$, $B = 0$, $C = 3$ works.

Convert the linear equation $y = 5x - 10$ to standard form, namely $Ax + By = C$. Solution: $5x - y = -10$.

Note that this is not the only solution. We could just as well multiply both sides of the equality by 2, to get $2y - 10x = -20$. The points satisfying the equality remain the same.

The same line can be represented by many equalities, obtained from one another by multiplying by a fixed number.

Parallel Lines

Look at the lines having the formulas $x + y = 1$ and $x + y = 3$.

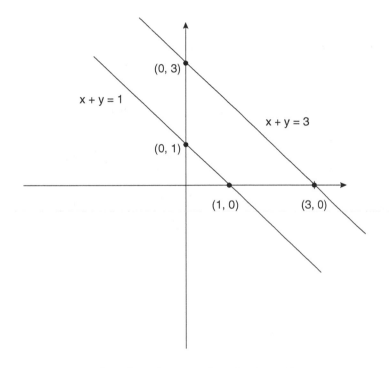

Drawing of $x + y = 1$ and $x + y = 3$

As can be seen from the drawing, the lines are parallel. In fact, this is clear. Moving terms, we can write the formulas as $y = -x + 1$ and $y = -x + 3$. This means that for every x the value of y in the second line is 2 more than in the first. So, the second line is obtained by shifting the first line 2 units up, so it is parallel to the first.

Similarly, the lines described by $3x + 4y = 5$ and $3x + 4y = 10$ are parallel, since moving terms transforms the first into $y = \frac{-3x}{4} + \frac{5}{4}$, and the second into $y = \frac{-3x}{4} + \frac{10}{4}$. So, the second is obtained by moving the first $\frac{10}{4} - \frac{5}{4} = \frac{5}{4}$ units up.

In general, $Ax + By = C$ and $Ax + By = D$ are expressions of parallel lines — or equal, if C = D.

Show that the lines $x + y = 10$ and $3x + 3y = 10$ are parallel.

(Hint: divide the second equality by 3.)

Two lines are parallel if and only one of them can be written as $Ax + By = C$ and the other as $Ax + By = D$, where $C \neq D$. For example, $2x + 3y = 0$ and $2x + 3y = 6$ are formulas of parallel lines.

Find pairs or parallel lines among: a. $3x + 2y = 10$ b. $3x + 9y = 10$ c. $x + 3y = 10$ d. $6x + 4y = 10$ e. $10x + 30y = 10$.

A Geometric Location of a Different Nature

Not all geometric locations are circles and lines. Here is an equality of another type: $|x| + |y| = 2$. What points (x, y) satisfy this condition? Note, first, that in such a point $|x|$ cannot exceed 2, since if it did, $|x| + |y|$ would be larger than 2 ($|y|$ being non-negative). Similarly, $|y|$ is at most 2.

To draw the figure, break into 4 cases. In the first, $x \geq 0$, $y \geq 0$. In such a case, $|x| = x$ and $|y| = y$, and the equality is then simply $x + y = 2$ — this is a straight line, in fact part of it, since as we said $x \leq 2$ and $y \leq 2$, so it is the part of the line between the points $(2, 0)$ and $(0, 2)$, like this:

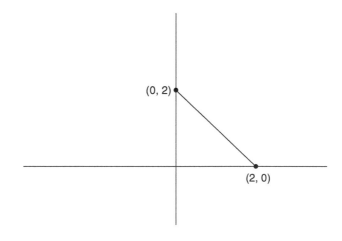

Drawing of part of the location

Another case to be considered is $x \geq 0$, $y \leq 0$. In this case the equality is $x - y = 2$.

Finish the other cases, to show that the location is a square:

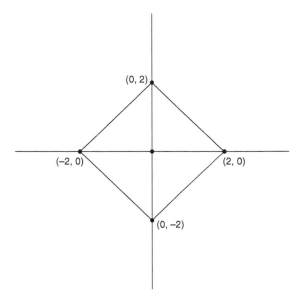

Drawing of the entire location $|x| + |y| = 2$

Why are the Axes Perpendicular?

Why did Descartes choose the axis to be perpendicular to each other? Could he have chosen them like this, for example?

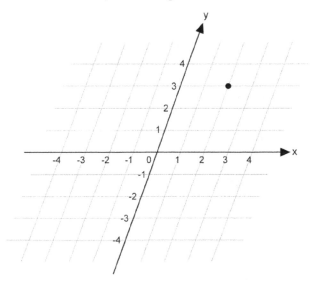

In principle, the answer is "yes". But life would be more complicated. Symmetry simplifies things. For example, the formula for the distance between points would be more complicated in skew coordinates.

Graphs

Connections

The Cartesian coordinate system is a way to visualize links.

Example: Reuben and Daniel earned together $100 doing a common job. Call x the amount Reuben got, and y the amount Daniel got. The connection between x and y is then $x + y = 100$.

The relation between the two numbers is described geometrically by the geometric location of the points (x, y) satisfying the condition. It is:

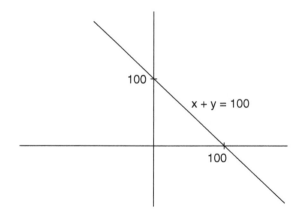

The points (x, y) satisfying $x + y = 100$

(We allowed negative numbers because one of them may be in debt.)

Graphs of Functions

Links are often given by functions, so the Cartesian coordinates allow graphic depiction of functions, indeed called "graphs". Here is its definition:

> The graph of the function $f(x)$ is the set of points (x, y) satisfying $y = f(x)$.

So, a graph is a geometric location, described by the equality $y = f(x)$.

Example: Let $f(x) = x + 1$. The graph of this function is the collection of points (x, y) satisfying $y = x + 1$. Here is the graph:

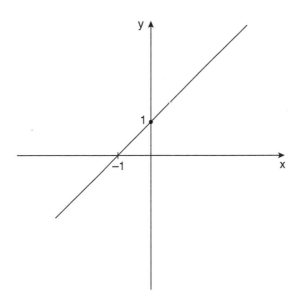

The graph of the function $f(x) = x + 1$

When drawing a graph of a function, you go along the x axis, and at every x the function tells you how high above the axis should you go. If $f(3) = 10$, you go at $x = 3$ to height 10.

For each of the following functions find 4 points on its graph, and then draw the graph, each on a separate coordinate system:

$$f(x) = x, \quad g(x) = 2x + 3, \quad h(x) = -x, \quad k(x) = \frac{1}{2}x.$$

Draw the graph of $f(x) = x^2$.

Start with some points on the graph:

$(0, 0^2) = (0, 0), \ (1, 1^2) = (1, 1), \ (2, 2^2) = (2, 4), \ (-2, (-2)^2) = (-2, 4).$

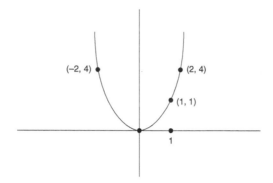

The graph of $f(x) = x^2$

Not Every Connection is a Function

Not every curve in the plane describes a function. For example, $x^2 + y^2 = 25$ is not the graph of a function, since for $x = 4$, for example, there are two values of y: both points $(4, 3)$ and $(4, -3)$ lie on the curve. A function should yield a single output for every input. For $x = 10$ there are no values at all — see drawing.

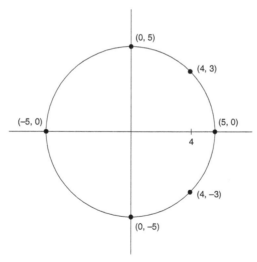

The circle is not the graph of a function

Shifting a Graph Vertically

Draw the graph of $g(x) = x^2 + 3$. Solution: this is just the graph of $f(x) = x^2$, shifted 3 units up. At every x, it is 3 units higher.

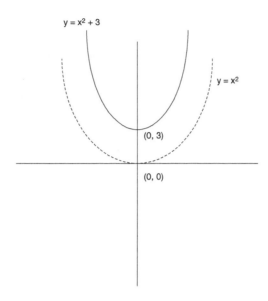

The graph of $f(x) = x^2$ dashed, above it the graph of $g(x) = x^2 + 3$

Slope and Rate of Change

Example: In the following drawing the graphs of $f(x) = 3x$ and $g(x) = x$ are drawn together.

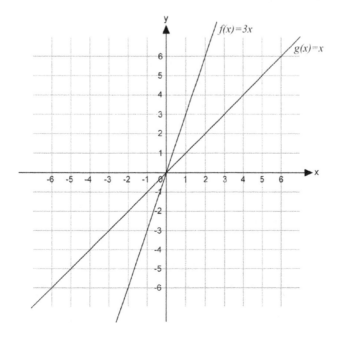

Think of each graph as a mountain, and imagine a mountaineer scaling it, from left to right. Which graph is steeper? Of course, that of $f(x)$. In fact, it is 3 times steeper, which can be defined mathematically as follows. Going on the graph $y = x$ of g, when you move one unit to the right you go up one unit. Going on the graph $y = 3x$ of f, when you move one unit to the right, you go up 3 units, three times more than in the graph of g.

But do you remember? This is the rate of change! "How much the function changes when the variable changes by 1" is precisely the rate of change. And this, we found, is the coefficient of x.

Definition: The slope of the graph of $f(x) = ax + b$ is its rate of change, which is a.

The graph of a linear function has constant slope. In non-linear functions, like in real life mountains, the slope changes from point to point.

> The price of gasoline is \$2 per liter. What is the rate of change of the price you pay, as a function of the number of liters you buy? Of course, 2. Or more precisely, 2 dollars per liter.
> You pump 10 liters a minute. What is the rate of change in the price the pump shows per minute?
>
> Answer: 10 times 2, which is 20 dollars per minute.

Units, like numbers, cancel out. If you pay $2\frac{dollars}{liter}$ and you pump $10\frac{liter}{minute}$, you pay $2\frac{dollars}{liter} \times 10\frac{liter}{minute} = 20\frac{dollars}{minute}$. The "liter" cancelled out.

The rate of change of a "function of a function" is the product of the rates of change of the two functions.

Ascending and Descending Functions

The function $f(x) = 3x$ has a positive change rate 3. This means that when x increases, the function grows. It grows 3-fold, but what we care of at this point is that it grows, positively. The function $g(x) = -x$, by contrast, has negative growth rate (-1), meaning that when x increases, the function decreases. A function with non-negative change rate is called "ascending", and a function with non-positive change rate is called "descending".

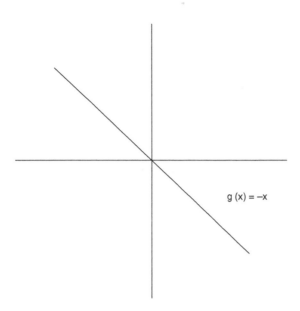

g (x) = –x

The function $f(x) = 3x$ is ascending, meaning that going right the graph goes up. The function $g(x) = -x$ is descending, meaning that its graph goes down.

In non-linear functions, there may be places where the function is ascending, and others where the function is descending. For example, the function $f(x) = x^2$ is descending for negative values of x (when you go with x from (-2) up to 0, the function descends from 4 to 0), and ascending for positive values of x.

A linear function $f(x) = ax+b$ is ascending if a is non-negative, descending if a is non-positive, and it is constant if $a = 0$.

Is the weight of a baby an ascending function of time, in her first year? (This is not a mathematical question. The surprising fact is that it is common for the weight to drop in the first weeks of life.)

Admonition: A Function is Not its Graph

Since Descartes' time, a function is automatically associated with its graph. And still, it is important to bear in mind that the graph is not the function, but a visualization of it. Confusing the two is like confusing a person with his or her photograph.

Unfortunately, the confusion often occurs, and its most common man-ifestation is in denoting the function by y. Sentences like "let $y = x^2 - x$ be a function" are often to be met in textbooks. This is misleading: the y coordinate is used to describe the value of the function, but it is the name of a coordinate, not a function. The correct formulation is: "let $f(x) = x^2 - x$ be a function and let $y = x^2 - x$ be its graph".

Functions Defined Separately in Different Domains

Sometimes a function obeys different rules for different values of the variable.

Example: A steamroller drives at a speed of 4 kmh for 3 hours. Afterwards it drives for 3 hours in the same direction at 2 kmh. Here is the graph of its location as a function of time:

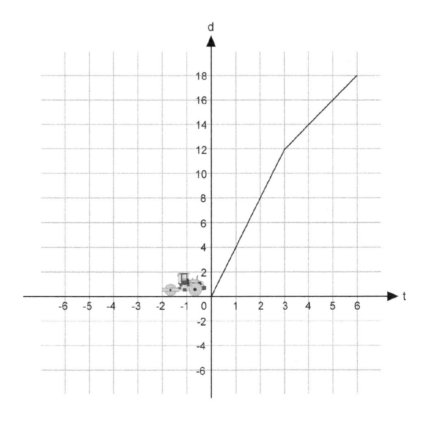

Deciphering Graphs

In this section we go in the opposite direction: from graph to function.

Example: A steamroller drives from Springfield north, and its distance from Springfield in the north direction at time t is depicted by the following graph. Describe its motion.

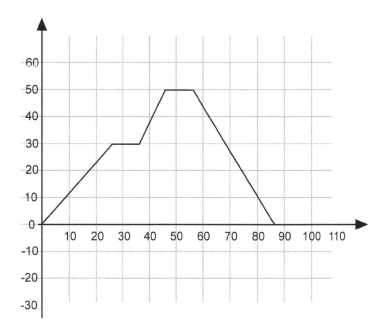

Solution: At $t = 0$ its distance from Springfield is 0, meaning that it is in Springfield.

  In the beginning, the roller goes at constant speed, which is expressed by the fact that the graph is straight. At $t = 4$, namely after 4 hours, $y = 12$, meaning that it is at distance 12 from Springfield. So, it speed was $\frac{12}{4} = 3$ kmh. As from $t = 4$ the graph is parallel to the t axis, meaning that its distance from Springfield is constant, it stopped for lunch. This lasted for 2 hours until $t = 6$. Then the distance starts decreasing, meaning that the roller is heading back to Springfield. In 2 hours (until $t = 8$) it went 12 km, so its speed is $\frac{12}{2} = 6$, and more precisely — (-6) kmh, since it went in the negative direction.

Describe the motion of a car whose distance from city A is given by the following graph:

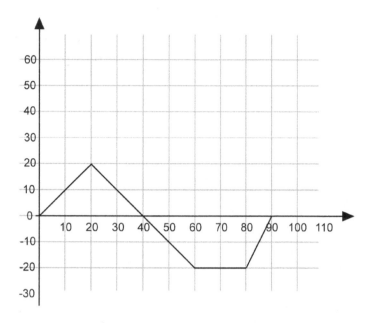

The time is measured in minutes, and the distance, north of the city, in km.

Lines, Once Again

The Tangent of an Angle

I want to view slopes more geometrically, and for this purpose a new notion is necessary: the "tangent of the angle". Look at the following right angle triangle:

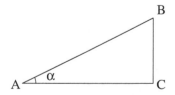

The tangent of the angle α (this is the Greek "a") is $\frac{CB}{AC}$, the ratio between the side opposite α and the side near α. It is denoted by $\tan \alpha$. A more appropriate notation would be $\tan(\alpha)$, because this is really a function of α, but $\tan \alpha$ is the more common notation.

Note that if we drew another triangle with the same angle α we would get the same ratio. Two right angle triangles, with the same angle α, share two angles (the right angle and α), so they share also the third angle (which in both of them is $180° - 90° - \alpha = 90° - \alpha$). This means that they are similar, meaning that they have the same ratios between sides, so in both the side opposite α and the side next to α is the same.

Calculate a. $\tan 0°$ b. $\tan 45°$.

(Answer to a: 0. Hint to b: show that the triangle in question is isosceles.)

Use diagram of triangle to show that
$$\tan(90° - \alpha) = \frac{1}{\tan \alpha}.$$

The rate of change of a linear function is the change in the function, divided by the change in the variable. In terms of the graph, it is the change in y divided by the change in x. Look at the following picture:

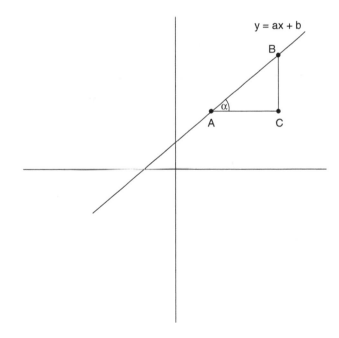

A and B are two points on the graph of the line $y = ax + b$. ACB is a right angle triangle. AC, the bottom side of the triangle, is the change in x when you go from A to B. CB is the change in y. So, the slope is $\frac{CB}{AC}$. But this is precisely $\tan \alpha$, for α the angle between the line and the x axis.

The Formula of a Straight Line, by Slope and a Point and by Two Points on the Line

A line can be represented in many ways. Here are three ways (not exhausting all).

1. A line $y = ax + b$ is known if we know a and b.
2. A line is known if we know a point on it and its slope.
3. A line is known if we know two points on it.

How do we translate one representation to another?

Example: Find the formula of a line going through the point $(3, 4)$ and having slope $\frac{1}{2}$.

Solution A: The formula of the line is $y = \frac{1}{2}x + b$, and we are looking for b. The fact that $(3, 4)$ is on the line means that it satisfies the equality, namely $4 = \frac{1}{2} \times 3 + b$. Solving for b gives $b = 4 - \frac{3}{2} = 2\frac{1}{2}$.

Solution B: Let (x, y) be a point on the line. Moving from $(3, 4)$ to (x, y) means going $x - 3$ in the x direction, and $y - 4$ in the y direction. The slope is the rate of change, which is the change in y divided by the change in x, which is $\frac{y-4}{x-3}$. Since the slope is $\frac{1}{2}$, we have $\frac{y-4}{x-3} = \frac{1}{2}$. Moving sides we get $y - 4 = \frac{1}{2}(x - 3) = \frac{1}{2}x - \frac{3}{2}$, and moving some more terms we get $y = \frac{1}{2}x + 2\frac{1}{2}$.

I find the second way more illuminating, since it clarifies the meaning of the slope.

Formula of a Line by Slope and a Point

What is the formula of a line going trough $(0, 0)$ and having slope 5?

Answer: $\frac{y-0}{x-0} = 5$, or otherwise written, $y = 5x$.

In general, a line with slope m passing through a point (x_0, y_0) (the index "0" is a way of saying "these are specific, fixed values of x and y") is:

$$\frac{y - y_0}{x - x_0} = m.$$

Moving sides:

$$y - y_0 = m(x - x_0).$$

What is the formula of a line going through $(-1, 2)$ and having slope 1?

(Solution: $\frac{y-1}{x+2} = 1$ — move sides to get to the more familiar form.)

A Line by Two Points on it

Through two points there goes one line. To find its formula, use the two points to calculate the slope.

Find the formula of the line going through the points $(1, 2)$ and $(3, 5)$.

Solution: We first find the slope. The change in y between the two points is $5 - 2$, and the change in x is $3 - 1$. So, the slope is:

$$\frac{5-2}{3-1} = \frac{3}{2}.$$

We now know the slope, and a point on the line (in fact, two points), $(1, 2)$. So, the formula of the line is

$$\frac{y-1}{x-2} = \frac{3}{2}.$$

Moving sides: $y - 1 = \frac{3}{2}(x - 2) = \frac{3}{2}x - 3$, or $y = \frac{3}{2}x - 2$.

In general, the formula for the line going through two points, say (x_1, y_1) and (x_2, y_2), (again, indices just say "these are fixed points") is:

$$\frac{y - y_1}{x - x_1} = \frac{y_2 - y_1}{x_2 - x_1}.$$

For each of the following pairs of conditions find the formula of the line satisfying them:

    a. Slope 3, going through $(0, 0)$
    b. Slope 0, going through $(8, 5)$
    c. Going through $(8, 5)$ and $(10, 5)$
    d. Going through $(8, 5)$ and $(8, 3)$

Find the formula of the line going through $(0, 0)$ and $(5, 7)$.

Solution: $\frac{y-0}{x-0} = \frac{7-0}{5-0}$, namely $\frac{y}{x} = \frac{7}{5}$, or $y = \frac{7}{5}x$.
Generalizing from this example:

> The slope of a line going through the origin and a point (c, d) is $\frac{d}{c}$.

Part 10

Arithmetic Sequences and Their Sums

The Average

On the average we are
1.5 meters tall.

What does the ant mean by "average"? There are two ways to say this. One is that the middle between its height and the height of the Giraffe is 1.5 meters. The other is that if we divide the sum of their heights between them, each gets 1.5 meters. The average between two numbers is their sum divided by 2.

The height of the ant is 0.25 cm. What is the height of the giraffe?

Answer: If the height of the ant was 0, the Giraffe would be 3m tall. To balance the addition of 0.25 cm, the height of the Giraffe should be 300–0.25 cm, namely $299\frac{3}{4}$ cm.

As noted, the middle between two number is their sum divided by 2. The following drawing explains why. The middle between 3 and 100 is the same as the middle between 103 and 0: we added 3 to 100, and subtracted 3 from 3. And the middle between 0 and 103 is half of 103, namely $\frac{100+3}{2}$.

0   3   100   103

The middle between 3 and 100 is the middle between 0 and 103

The average of two numbers a and b is their middle, which is $\frac{a+b}{2}$.

Another Point of View

What is the middle between 80 and 100?

I am sure that you found the answer, 90, not by adding 100 and 80 and dividing by 2, but by going from 80 "half way", namely adding to it 10, half of $100 - 80$.

Doing this for general numbers will bring us to the same formula for the middle obtained above. Let a and b be two numbers. Their middle is obtained by going half way from a, and half of the distance between the numbers is $\frac{1}{2}(b - a)$. We get that the middle is:

$$a + \frac{1}{2}(b - a) = a + \frac{1}{2}b - \frac{1}{2}a = \left(a - \frac{1}{2}a\right) + \frac{1}{2}b = \frac{1}{2}a + \frac{1}{2}b = \frac{a + b}{2}.$$

Find the middle between: a. 0 and x; b. 100 and 100; c. 97 and 100; d. $x - 3$ and $x + 3$.

What is the middle between the height of the Everest, which is 8882 meters high, and the Dead Sea, which is (-420) meters high?

The Average of Many Numbers

The average income of the inhabitants of a country is the total income, divided by the number of people. The average of 3 numbers a, b and c is $\frac{a+b+c}{3}$.

The average of n numbers is their sum divided by n.

---

30 students of a class collected planks for a bonfire. 20 children collected 5 planks each, one child collected 80 planks, and the rest did not collect any. How many planks on the average did each collect?

Answer: Together they collected $(20 \times 5) + 80 = 180$ planks, so on the average they collected $180 : 30 = 6$ planks.

Write an expression for the average of 4 numbers, a, b, c and d.
The highest income of a state employee is $150,000 a month, and the lowest income is $2000. What is the middle between the two numbers? Is it the same as the average income of state employees? If not, explain why.

(Hint: very few get salaries close to $150,000 a month.)

In a class of 30 students the grade of one student was raised by 10 points. How much did the average grade rise?

(Answer: $\frac{1}{3}$ of a point.)

A Point of Reference

Jane got 100 in 3 exams, and in the fourth she got 99. What was her average grade?

Answer: Take 99 as a point of reference. Relatively to 99, she got 1 in 3 exams, and 0 in the fourth. The average of these is $\frac{1+1+1+0}{4} = \frac{3}{4}$, so her average was $99 + \frac{1+1+1+0}{4} = 99\frac{3}{4}$.

The principle used here is that adding the same number a to each of a bunch of numbers adds a to their average. In the above case, a was 99.

> Calculate the average between 0,2 and 3. Then calculate the average between 1000, 1002 and 1003.
> What happens to the average of some numbers if each of them is multiplied by 10? What is the average of 1, 2 and 3? What is the average of 10, 20 and 30?

Finding the Sum by the Average

Example: 30 students collected 6.1 planks on the average. How many planks did they collect?

Answer: $30 \times 6.1 = 183$ planks.

The reason: the average is the sum divided by 30, so the sum is the average multiplied by 30.

> The sum of n numbers is their average multiplied by n.

Free Fall

In my opinion, not enough physics is taught in middle school. I want to make here a little contribution in this direction, by discussing a physical topic, "free fall". It is connected to the notion of the average, and to another related topic — the sum of arithmetic series. But before I start, I will ask you to make a small guess:

> Try to guess how much time will it take for a body to fall 50 km from an airplane? (Of course, such height is more that of a spacecraft than that of a plane...)

Imagine a tall building from which two stones are dropped, one 1 kg heavy, and the other of weight 10 kg. Which stone will reach the ground first? The ancient Greeks believed that the heavy stone will reach the ground

10 times faster. The legend goes that to refute this, Galileo dropped stones from the Pisa tower. You need to drop from a high point because in short distances the fall is so quick that it is hard to measure. But Galileo solved this problem in a cleverer way: he used slanted slopes, on which rolled balls. On a slanted slope the fall is slow enough to follow. He discovered that heavy bodies fall just as fast as light bodies. Later Newton explained why this is so.

Constant Acceleration

What Galileo found, and Newton explained, was that all bodies, heavy or light, fall with constant acceleration. Every second the speed of a falling body grows by about 9.8 meters per second. The number 9.8 is a bit inconvenient, so I will follow a custom, and round it to 10.

So, when a body starts falling from a tall building it starts at the speed of 0 meters per second (m/s). After one second its speed is 10 m/s. After 2 seconds its speed is 20 m/s. After 3 seconds its speed is 30 m/s, and so on. Every second the speed goes up by 10 m/s. So, after t seconds the speed is $10t$ m/s.

On the moon, for example, the acceleration is smaller — about 2 m/s each second. But still, it is the same for heavy body and for light bodies.

Newton's Explanation

Why is the acceleration independent of the weight of the body? Newton understood the secret. He realized two things:

1. A body A 10 times as heavy as a body B is attracted to the center of the earth by a force 10 times larger than B.
2. Inducing the same acceleration on A as on B requires force 10 times larger.

In short — the force of gravitation on A is larger, but accelerating a heavy body requires larger force. The two factors balance each other.

Average Speed and Distance

Let us now use this knowledge to find the distance traveled during the first 20 seconds. The starting speed is 0 m/s. After 20 seconds, the speed is 10×20 m/s, namely 200 m/s. Since the acceleration is constant, the average speed during this time is the middle between 0 and 200, which is 100 m/s.

And if you travel for 20 seconds at an average speed of $100\,\text{m/s}$, you travel $20 \times 100 = 2000$ meters.

Let us now find the general formula — what is the distance the body covers in t seconds. At the starting point, its speed is $0\,\text{m/s}$. After t seconds the speed is $10t\,\text{m/s}$. So, the average speed is the middle between the two — $5t\,\text{m/s}$. And going at an average speed of $5t\,\text{m/s}$ for t seconds, the body covers $t \times 5t = 5t^2$ meters.

What distance does a body fall during the first second of its fall?

Putting $t = 100$ in the formula, we see that in 100 seconds a body falls $5 \times 10,000 = 50,000$ meters. This is the answer to the little guess I asked you to make in the beginning of this chapter: it takes 100 seconds, less than two minutes, to fall $50\,\text{km}$! A bit surprising, no?

How much time does it take a body to fall 500 meters? Answer: the equation is $5t^2 = 500$, so $t^2 = 100$, or $t = 10$. It takes the body 10 seconds to fall 500 meters.

(Of course, there is another solution, $t = -20$. It corresponds to the possibility that 20 seconds ago somebody threw the body up, so that when the body reaches its peak its speed is zero.)

What distance does a falling body cover during the 7-th second of its fall?

First solution: the speed of the body at $t = 6$ is $60\,\text{m/s}$, and its speed at $t = 7$ is $70\,\text{m/s}$. Its average speed during the 7-th second is therefore $65\,\text{m/s}$. The distance it passes during this second is therefore 65 meters.

Second solution: by $t = 6$ the body has gone 5×6^2 meters, and by $t = 7$ the body has gone 5×7^2 meters. During the 7-th second it went therefore $5 \times 7^2 - 5 \times 6^2 = 5 \times (7^2 - 6^2) = 5 \times 13 = 65$ meters.

Follow this example to show that between time u and time v the body goes $5(v - u)(u + v) = 5(v^2 - u^2)$ meters. What happens when you put in this formula $u = 0$? Is this a familiar formula?

Arithmetic Sequences

The sequence of numbers 3, 7, 11, 15, 19 is called an "arithmetic sequence". It contains 5 terms. There are also infinite arithmetic sequences, like 3, 7, 11, 15, 19... The characteristic of an arithmetic sequence is that at every step it increases, or decreases, by a constant number, which is called the "difference" of the sequence, and is usually denoted by d. In the above example, $d = 4$. The sequence 11, 9, 7, 5, 3, 1, -1, -3 is also arithmetic, with difference $d = -2$, since it "increases" by (-2) at each step, meaning that it decreases by 2. The constant sequence 5, 5, 5, ... is also arithmetic, with difference 0.

> For each of the following sequences determine whether it is arithmetic, and if so find its difference. If it is not arithmetic, try to find the regularity and the next element:
>
> a. 1, 3, 6, 10, 15, 21   b. 0, 1, 0, 1, 0, 1   c. 5, 4, 3, 2, 1
> d. 1, 2, 3, 4, 5   e. 1, 2, 6, 24, 120, 720 f. 1, 4, 12, 32, 80.

(Hint for the last sequence: divide the first term by 1, the second term by 2, the third term by 3, and so on. What did you get?)

The Sum of an Arithmetic Sequence

One of the best known stories in the history of mathematics is about how the famous German mathematician Karl Friedrich Gauss discovered a method for summing up the terms of an arithmetic sequence at the age of 7. He was not the first to make this discovery, but for a 7 years old boy it is indeed a magnificent insight.

The story is that one day his teacher wanted to have a rest, so he asked his students to sum up all numbers between 1 and 100, namely $1+2+3+\cdots+100$. To his surprise, little Karl Friedrich came with the answer in a few seconds.

Before proceeding to read, please try to guess the sum — can you estimate it? Is it more than 100 times 100, or less?

A famous Hugarian educator always asks his students to estimate the answers to numerical questions, before doing the precise calculation. This gives a good feel for the problem, and may sometimes even lead to a quick solution. For one thing, it frees the student of anxiety, since when first approximation is in question, every estimate is allowed.

The trick ascribed to Gauss is this: he paired the numbers up like this: $1 + 100$, then $2 + 99$, then $3 + 98$ until $50 + 51$. Each of these pairs sum up to 101, and there are 50 pairs, so the sum is 50×101, which is 5050.

In general, to calculate the sum $1 + 2 + \cdots + (n - 1) + n$, pair 1 with n, 2 with $n - 1$, 3 with $n - 2$, and so on. Each of these pairs sums up to $n + 1$, and there are $\frac{n}{2}$ pairs. So, we get:

$$1 + 2 + 3 + \cdots + n = \frac{n}{2} \times (n+1) = \frac{n(n+1)}{2}.$$

Using the Average

There is a problem with the above calculation, which is that n needs not be even. If it is odd, the numbers do not pair up. This is easily solved by adding 0 at the beginning, which of course does not change the sum, but makes the number of terms even. But there is an even more elegant solution, using the notion of the average. The average of the numbers between 1 and n is the middle between 1 and n, which is $\frac{n+1}{2}$. The sum of n numbers whose average is $\frac{n+1}{2}$, so we know, is $\frac{n+1}{2} \times n = \frac{n(n+1)}{2}$ — the same formula we found before.

The Sum of General Arithmetic Sequences

The property of the sequence $1, 2, \ldots, n$ that was used here to calculate its sum was its being "evenly paced". This is why its average is the middle between its first and last element. But being "evenly paced" means just being an arithmetic sequence.

> In each of the two following sums write the difference of the sequence, the number of summands, the average and the sum:
>
> a. $3 + 7 + 11 + 15 + \cdots + 99$
> b. $100 + 102 + 104 + \cdots + 200$.

(The difference is 4; the number of summands is $\frac{99-3}{4} + 1 = 25$, since the sequence partitions the interval between 3 and 99 into $\frac{99-3}{4} = 24$ intervals of length 4, to which the first point, 3, should also be added; the average is $\frac{99+3}{2} = 51$; the sum is $25 \times 51 = 1275$.)

And now, the sum of a general arithmetic sequence. An arithmetic sequence is determined by 3 parameters: its first element a, its difference d and the number of its elements n. The sequence is then:

$$a, a + d, a + 2d, a + 3d, \ldots, a + (n-1)d.$$

Why is the last term $a + (n-1)d$? Because the first term is $a = a + 0d$, so the multiples of d go from 0 to $n - 1$, which gives n terms.

The average of this sequence is the middle between the first and the last terms, namely $\frac{a+a+(n-1)d}{2} = a + \frac{n-1}{2}d$. So, the sum of the sequence is: $n \times \left(a + \frac{n-1}{2}d\right)$.

Find the sum of all multiples of 7 between 1 and 700. Do it using the above formula, and also using the sum $1 + 2 + 3 + \cdots + 100$.

Find the sum $-100 + (-92) + (-84) + \cdots + 84 + 92 + 100$, using the above formula. Is there an easier way?

The Connection to Free Fall

In a free fall, the speed of the falling body after 1 second is 10 m/s, after 2 seconds 20 m/s, after 3 seconds 30 m/s, and so on. This is an arithmetic sequence. The rate of change of the speed is constant, so the average speed is the middle between the initial speed and the final speed, just like in an arithmetic sequence. The distance is the sum of distances traveled in second 1, second 2, and so on — just as in the sum of an arithmetic sequence. And indeed, the formula for the sum $1 + 2 + 3 + \cdots + n = \frac{1}{2}n(n+1)$, is similar to the formula for the distance traveled by a falling body in the first t seconds: $\frac{1}{2} \times 10 \times t^2$.

Part 11

Polynomials

Definition

The expression $3x^2 + 5x + 7$ is called a "polynomial in the variable x". "Poly" means in Greek "many", like in "polygamy" (many wives). The origin of the name "polynomial" is that it is the sum of many terms. But these are special terms: each of them is a number times a power of the variable.

A polynomial in the variable x is also a function, and hence we shall use function notation, often calling a polynomial $p(x)$.

Calculate $p(1)$ for the polynomial $p(x) = 3x^2 + 5x + 7$.

Coefficients and Degree

The numbers 3, 5 and 7 are called the "coefficients" of the polynomial $3x^2 + 5x + 7$. The highest power term is called the "leading term", and its coefficient is called the "leading coefficient". Its power is called the "degree" of the polynomial. The degree of a polynomial $p(x)$ is denoted by $\deg p(x)$ or $\deg(p(x))$. For example, the leading term in $3x^2 + 5x + 7$ is $3x^2$, the leading coefficient is 3, and the degree of the polynomial is 2, written $\deg(3x^2 + 5x + 7) = 2$. The importance of the degree is that the highest degree term is the significant one for large x. Putting $x = 1000$ in the polynomial $3x^2 + 5x + 7$, the significant term is $3x^2$, which for $x = 1000$ is 3 million. The term not containing x is called the "free term". The free term in $3x^2 + 5x + 7$ is 7.

It is customary to write the terms of the polynomial in descending order of power. But this is not compulsory — instead of $3x^2 + 5x + 7$ we could also write $5x + 7 + 3x^2$.

Write the degree, the leading terms and the free terms in the following polynomials: a. $1 + 2x^3$ b. 1 c. 10 d. $x^2 + x^2 - 3$ e. x^2

(Hint: in d first combine the highest terms to one term.)

What is $p(0)$ for any polynomial?

(Answer: the free term.)

There is one special polynomial which has no leading term: 0. Its degree, for reasons to be understood later, is defined as minus infinity.

A Few Familiar Examples

a. A constant number, say 7, is a polynomial of degree 0. It is equal to $7 = 7 \times x^0$ (remember that $x^0 = 1$).

b. A linear expression $ax + b$ is a polynomial of degree 1, since it can be written as $ax^1 + bx^0$.

c. The most familiar use of polynomials is in the decimal system. A decimal representation of a number is nothing but a polynomial, substituting 10 for its variable. For example, $357 = 3 \times 10^2 + 5 \times 10 + 7$, which results from substituting $x = 10$ in the polynomial $3x^2 + 5x + 7$. In the polynomials appearing in the decimal system, all coefficients are at most 9.

Adding and Subtracting Polynomials

The sum of two polynomials is a polynomial, and so is their difference. For example:

$$(3x^2 + 5x + 7) + (x^2 + 1) = (3x^2 + x^2) + 5x + (7 + 1) = 4x^2 + 5x + 8, \text{ and}$$

$$(3x^2 + 5x + 7) - (x^2 + 1) = (3x^2 - x^2) + 5x + (7 - 1) = 2x^2 + 5x + 6.$$

What is the coefficient of x^2 in the sum of the two polynomials $3x^2 + 5x + 7$ and $x^5 + 10x^2 + 1$?

(Answer: 13.)

Find two polynomials of degree 3 whose sum is of degree 1.

(One solution: $x^3 + x^2 + x + 1$ and $-x^3 - x^2 + x + 1$.)

What can you say about the degree of the difference between a polynomial of degree 4 and a polynomial of degree 3?

(Answer: the difference is of degree 4. Nothing will cancel out the leading term in the degree 4 polynomial.)

Multiplying Polynomials

Polynomials are multiplied using the distributive law. Example:

$$(3x^2 + 5x + 7)(2x^3 + x + 4) = 3 \cdot 2x^5 + (5 + 3 \cdot 4)x^2 + (5 \cdot 4 + 7)x + 7 \cdot 4$$

$$= 6x^5 + 17x^2 + 27x + 28$$

It is instructive to follow the opening up of the brackets. You realize that:

a. The degree of the product is the sum of the degrees of the multiplicands. In the example: $5 = 2 + 3$. In fact, the leading term of the product is the product of the leading terms in the multiplicands. In the example: $3x^2 \times 2x^3 = 6x^5$. There is no other term that can cancel out this product.

In a formula: $\deg p(x)q(x) = \deg p(x) + \deg q(x)$.

b. The free term in the product is the product of the free terms in the multiplicands. In the example: $7 \times 4 = 28$. Again, nothing is going to cancel this out.

c. Other terms in the product may be produced by more than one term. In the example above, x^2 is obtained in the product both from $3x^2 \times 4$ and $5x \times x$.

(*) Explain why the rule $\deg p(x)q(x) = \deg p(x) + \deg q(x)$ is true also when one of the polynomials (say $p(x)$) is 0. (This is where the definition of $\deg(0) = -\infty$, namely minus infinity, is needed. Note: $-\infty + \deg q(x) = -\infty$, because minus infinity is a "black hole", absorbing any number. Minus infinity plus any number is minus infinity.)

Comparing Polynomials

What is larger: $2x + 3$ or $3x + 2$? Hard to tell, because they have the same degree. But we can order the polynomials by degree: a polynomial of degree 3 is "larger" than a polynomial of degree 2.

The chase after roots of polynomials

Already the ancient Chinese could solve quadratic equations, like $3x^2 + 5x + 7 = 5$. "Quadratic" means that the polynomial appearing in them is of degree 2. Moving terms from the right side to the left side, we can assume that the right hand side is zero. For example, the above equation can be written as $3x^2 + 5x + 2 = 0$. We say that a number a is the root of a polynomial $p(x)$ if $p(a) = 0$. For example, (-1) is a root of the above equation, since $3(-1)^2 + 5 \times (-1) + 2 = 0$.

Solving equations of degree higher than 2 was the holy grail for mathematicians for many centuries. In the 15-th century mathematicians found formulas for solving equations of degrees 3 and 4. In the beginning of the 19-th century something very surprising was discovered: there is no formula for solving equations of degree 5 and more. The discoverer of this fact was Niels Abel, a Norwegian mathematician who was born in 1802, and died at the age of 27 in great poverty.

Dividing Polynomials

In middle school division of polynomials is rarely taught. I think this is a mistake — polynomial division is both instructive and important. And it is not very hard, for those who understand long division of numbers.

How to Write Division with Remainder

Let us recall the three ways of writing division with remainder, mentioned in "to go back or not to go back". The fact that $37 : 5 = 7$ with remainder 2 can be written as:

a. $37 : 5 = 7(2)$
b. $37 = 7 \times 5 + 2$
c. $\dfrac{37}{5} = 7 + \dfrac{2}{5}$

The point is that the remainder 2 is smaller than the divisor 5 (otherwise we could continue to divide it). This is true also when dividing polynomials, where "smaller" is having smaller degree.

A Toy Example: Dividing by the Polynomial x

Dividing a polynomial by a number is easy. For example, $(x^2 + x + 1) : 3 = \frac{x^2}{3} + \frac{x}{3} + \frac{1}{3}$. So, division becomes interesting only when the divisor is a real polynomial, containing the variable. And the simplest of those is the polynomial x.

Examples:

1. (no remainder): $(x^2 + 2x) : x = x + 2$
2. $(x^2 + 2x + 5) : x = x + 2(5)$, which can also be written as:

$$x^2 + 2x + 5 = (x + 2)x + 5, \quad \text{or} \quad (x^2 + 2x + 5) : x = x + 2 + \frac{5}{x}.$$

From this example you can see: the remainder in a division $p(x) : x$ is the free term in $p(x)$. And this term is as recalled $p(0)$, namely the value of $p(x)$ when $x = 0$.

> What is the remainder in the division $(x^5 + x + 7) : x$?
> Can you determine by $p(0)$ whether the polynomial $p(x)$ is divisible by x?
>
> (Answer: $p(0)$ should be 0.)

Three More Simple Examples

Often polynomial division can be performed using simple tricks. Here are three examples:

a. $x : (x+1)$. Write $\frac{x}{x+1} = \frac{(x+1)-1}{x+1} = 1 - \frac{1}{x+1}$. The quotient is 1, and the remainder is (-1).

b. $x^2 : (x+1)$. We shall use the identity $x^2 - 1 = (x+1)(x-1)$. Write: $\frac{x^2}{x+1} = \frac{(x^2-1)+1}{x+1} = \frac{(x+1)(x-1)+1}{x+1} = \frac{(x+1)(x-1)}{x+1} + \frac{1}{x+1} = x - 1 + \frac{1}{x+1}$. The quotient is $x - 1$, and the remainder is 1.

c. $x^2 : (x+5)$. We shall use the identity $x^2 - 25 = (x-5)(x+5)$. This can be written as: $x^2 = (x-5)(x+5) + 25$, which is one way of writing that $x^2 : (x+5) = (x-5)$ with remainder 25.

Note that in all three examples the degree of the remainder is 0, smaller than the degree of the divisor, which is 1.

Use the identity $x^2 = (x-3)(x+3) + 9$ to calculate $x^2 : (x-3)$.

A Systematic Way of Dividing by a Linear Polynomial

Tricks are nice, but systematic methods are better. There is a special type of polynomials that dividing by them can be done in an easy way. These are polynomials of the form $x - a$, like $x - 2$ (here $a = 2$) or $x + 5$ (here $a = -5$). The way to do this is substitute $x - a + a$ for x.

Divide $(2x^2 + 1) : (x + 5)$.

Solution: write $x = (x + 5) - 5$. Then $2x^2 + 1 = 2[(x+5) - 5]^2 + 1 = 2[(x+5)^2 - 2 \times 5 \times (x+5) + 25] + 1 = 2(x+5)^2 - 20(x+5) + 51$, and hence $(2x^2 + 1) : (x + 5) = [2(x+5)^2 - 20(x+5) + 51] : (x + 5) = 2(x+5) - 20$ with remainder 51.

This is possible to do in any division by a polynomial of the form $x - a$. If we wish to divide by a general linear polynomial, like $3x + 2$, we write $3x + 2 = 3\left(x + \frac{2}{3}\right)$, and then dividing by $3x + 2$ can be done in two steps: dividing by 3, and then dividing by $x + \frac{2}{3}$. The latter we have just seen how to do.

This shows that dividing by a linear polynomial can be done, with remainder of degree 0, namely a number. A surprising fact is that we can know the remainder even without performing the division!

The Remainder Theorem

Theorem: the remainder in the division $p(x) : (x - a)$ is $p(a)$.

Examples:

a. In the division $(2x^2 + 1) : (x + 5)$ we have $p(x) = 2x^2 + 1$ and $a = -5$. So, by the theorem, the remainder is $2 \times (-5)^2 + 1 = 2 \times 25 + 1 = 51$, just as we obtained in the calculation above.

b. For $a = 0$ the division is $p(x) : x$, and the theorem says that the remainder is $p(0)$. We saw that before.

Proof of the remainder theorem: We know that the remainder is a number. Call it r. Then, by one of the ways of writing remainders,

$$p(x) = q(x)(x - a) + r.$$

Put in this equality $x = a$. Then $x - a = 0$, and hence on the right hand side we get just r. On the left hand side we have $p(a)$. So, we got $p(a) = r$, namely the remainder is $p(a)$.

This theorem has a very useful corollary (consequence).

Corollary: $p(x)$ is divisible by $x - a$ if and only if $p(a) = 0$.

This is true simply because $p(x)$ being divisible by $x - a$ means that the remainder in the division is 0.

$p(x) = x^2 - 5x + 6$ satisfies $p(2) = 0$, hence $p(x)$ is divisible by $x - 2$. Indeed $x^2 - 5x + 6 = (x - 2)(x - 3)$. What is the other value of x for which $p(x) = 0$?

In each of the following divisions write the remainder, without performing the division. Also determine when there is divisibility with no remainder:

a. $(x^{10} - 1) : (x - 1)$ b. $(x - 1) : (x - 1)$ c. $(x^n - 1) : (x - 1)$ (n being any natural number) d. $(x^{10} - 1) : (x + 1)$ e. $(x + 1) : (x - 1)$ f. $(x^2 + x + 1) : (x - 2)$ g. $(x^2 - 5x + 6) : (x + 2)$.

Solution to a: 0, meaning that $x^{10} - 1$ is divisible by $x - 1$.

Explain using the remainder theorem why $p(x) = x^2 - 1$ is divisible by $x - (-1) = x + 1$. Find the quotient.

Answer: $p(x) = x^2 - 1$. We have $p(1) = 1^2 - 1 = 0$, so $p(x)$ is divisible by $x - 1$. But also $p(-1) = (-1)^2 - 1 = 1 - 1 = 0$, so $p(x)$ is divisible by $x + 1$. Indeed

$$p(x) = (x - 1)(x + 1).$$

The General Algorithm for Dividing Polynomials

There is a systematic way of dividing any polynomial by any polynomial. It is just a polynomial version of the familiar long division. I will exemplify it again in a simple case — dividing by a linear polynomial.

Calculate $(2x^2 + 1) : (x - 5)$.

We start by dividing leading terms. Namely, $2x^2 : x$. The result is $2x$. Think of the division as dividing $2x^2 + 1$ candies among $x - 5$ children. If the result is $2x$, then every child gets $2x$ candies, so the $x - 5$ children get $(x - 5) \times 2x$ candies, namely $2x^2 - 10x$. This is not the entire amount of candies we are trying to divide: we want to divide $2x^2 + 1$ candies. So, how many haven't we divided? $2x^2 + 1 - (2x^2 - 10x)$, namely $10x + 1$. This is what remains in our hands, having given each child $2x$ candies. And this we shall have to continue to divide. All this is written as in long division:

$$\begin{array}{l} 2x \\ \overline{} \\ 2x^2 + 1 \,|\, x - 5 \\ 2x^2 - 10x \\ \overline{} \\ 10x + 1 \end{array}$$

In the next step, we divide the remaining $10x + 1$ among the $x - 5$ children. So, we want to perform $(10x + 1) : (x - 5)$. Again, we divide leading terms, to get $10x : x = 10$. If each of the $x - 5$ children got 10 candies, we have distributed $10 \times (x - 5) = 10x - 50$ candies out of the $10x + 1$ we are supposed to distribute. There remain $10x + 1 - (10x - 50)$, namely 51 undistributed candies. This we cannot divide, since it is a number, which is of degree 0, while the divisor $x - 5$ is of degree 1. We write it all like this:

$$2x + 10 \quad (51)$$

$$\overline{}$$

$$2x^2 + 1 \,|\, x - 5$$
$$2x^2 - 10x$$

$$\overline{}$$

$$10x + 1$$
$$10x - 50$$

$$\overline{}$$

$$51$$

We can write the result of the division as: $\frac{2x^2+1}{x-5} = 2x + 10 + \frac{51}{x-5}$. The quotient is $2x + 10$ and the remainder 51.

Divide: a. $(2x^3 + 1) : (x - 5)$ b. $\frac{x^3}{x+1}$ c. $\frac{2x^3+1}{x+1}$.

Part 12

Equations with Two Unknowns

Two Unknowns Demand Two Pieces of Data

Suppose that you wish to unveil the identity of two persons, A and B, and that you are given just one piece of information connecting them, say that A is the father of B. This is clearly not enough. There are many pairs of father-son. You need another piece of information, like their residence. There is a good chance that in one prescribed apartment there is only one pair of father-son, so this information will suffice.

This is true also for numbers. To find one number, you need one piece of information. To find two numbers, you need two. For example, knowing that the sum of the ages of Bob and Alice is 10 leaves lots of possibilities open. Possibly Alice is 6 years old and Bob is 4, and possibly Alice is 15 and Bob is (-5) (meaning he will be born in 5 years), and there are infinitely more possibilities. But if you are told also the difference between the ages, you can find them:

> The sum of the ages of Bob and Alice is 10, and Alice is 4 years older than Bob. What are their ages?

Of course, Alice is 7 and Bob is 3.

Two Equations

Calling the age of Alice x and the age of Bob y, the two sets of data given in the above problem are:

$$x + y = 10$$

$$x - y = 4$$

This is an example of a system of two equations, in two unknowns. In this book we shall only deal with systems of linear equations, meaning that each unknown can appear only in degree 1, times a number. The unknowns cannot be multiplied by one another, or appear in a denominator. To give an idea of how a system of non-linear equations may look like, here is an example:

$$x^2 + y = 5$$

$$xy = 2$$

175

Can you guess a pair of values for x and y that solve these equations? (Try small numbers.)

Pairs of Numbers by Their Sum and Difference

Let us start with a trivial example:

Gerald and Harry have together $100, and both have the same amount. How much does each of them have?

Of course, $50.

Now let us change the problem a bit:

Gerald and Harry have together $100, and Gerald has $2 more than Harry. How much does each of them have?

Answer: Divide the sum equally between them, and then let Harry give Gerald 1 dollar. Now Gerald has 2 more than Harry. So the answer is, Gerald has 51, and Harry 49.

And what if Gerald has $24 more than Harry? Answer: Give Gerald and Harry $50 each, and then let Harry give Gerald $12. Gerald has 62, and Harry 48.

We have just solved the system $\begin{cases} x+y = 100 \\ x-y = 24 \end{cases}$. The solution is $x = 62$, $y = 48$.

Solve: $\begin{cases} x+y = 80 \\ x-y = 10 \end{cases}$, $\begin{cases} x+y = 97 \\ x-y = 11 \end{cases}$.

Let us now do it more systematically, in a way that will work in any pair of equations in two unknowns.

Equalizing Coefficients

Step 1: When adding makes no difference

Solve: $\begin{matrix} x+y = 10 \\ x+5y = 10 \end{matrix}$.

Something strange is happening here: adding y is the same as adding $5y$. This means that $5y = y$, implying that $y = 0$. To finish the solution, put $y = 0$ in one of the equations, say the first, to get $x+0 = 10$, namely $x = 10$.

Solve a. $\begin{matrix} 2x+y = 10 \\ 2x-y = 10 \end{matrix}$ b. $\begin{matrix} x+y = 10 \\ 2x+y = 10 \end{matrix}$.

Step 2: Find the difference

In a second type of equations, adding more (or less) of one variable does make a difference.

Solve: $\begin{array}{l} x+y=10 \\ x+5y=58 \end{array}$.

Solution: In the second equation there are 4 more y's than in the first, and this should account for the addition of 48 (the difference between 58 and 10), so $4y = 48$, meaning that $y = 12$. Plugging this in the first equation gives $x + 12 = 10$, that is $x = -2$. (What do you think will happen if we plug $y = 12$ in the second equation? Try!)

Solve a. $\begin{array}{l} x+2y=10 \\ x+5y=58 \end{array}$ b. $\begin{array}{l} x-2y=10 \\ x+5y=73 \end{array}$.

Step 3: Making the coefficients of one variable equal

The secret in the previous examples was that one variable appeared in both equations with the same coefficient. If you don't have this, it is just possible to make it happen! All you have to do is multiply both sides of one of the equations by the appropriate number.

Solve: $\begin{array}{l} 2x+3y=18 \\ x-y=-1 \end{array}$.

Solution: Multiply the second equation by 2, to get: $\begin{array}{l} 2x+3y=18 \\ 2x-2y=-2 \end{array}$. In the first equation there are $5y$ more than in the second, which should account for the difference $18 - (-2) = 20$. So $5y = 20$, or $y = 4$. Putting this in the first equation gives $2x + 12 = 18$, or $x = 3$.

Step 4: A more systematic approach — subtracting one equation from the other

It is customary to do the last stage in the solution in a more systematic way. We got to the system

$$2x + 3y = 18$$

$$2x - 2y = -2$$

Subtract the second equation from the first, meaning that we subtract each side of the second equation from the same side in the first. The first equation becomes then:

$$2x + 3y - (2x - 2y) = 18 - (-2)$$

Namely: $5y = 20$, or $y = 4$.

The subtraction of sides method has the advantage that it can be done automatically, without thinking.

Solve the system using this way $\begin{array}{l} x + y = 10 \\ x - y = 4 \end{array}$.

Step 5 (and last): Multiplying both equations by numbers

Sometimes it is not so simple to equalize coefficients. For example:

Solve: $\begin{array}{l} 2x + 3y = 13 \\ 3x + 7y = 22 \end{array}$.

We can multiply the first equation by $\frac{3}{2}$, and then the system becomes:

$$3x + \frac{9}{2}y = \frac{39}{2}$$

$$3x + 7y = 22$$

This is OK for those who love fractions. For the lazier, there is another way out: multiplying the first equation by 3, and the second equation by 2. We get:

$$6x + 9y = 39$$

$$6x + 14y = 44$$

Now the coefficients of x are the same in both equations. Subtracting the second equation from the first gives: $6x + 9y - (6x + 14y) = 39 - 44$, which upon collecting terms becomes $-5y = -5$, yielding $y = 1$. Putting this in the first equation gives $6x + 9 = 39$, or $6x = 30$, meaning that $x = 5$.

Solve $\begin{array}{l} 7x + 3y = 19 \\ 5x + 7y = -1 \end{array}$.

(Hint: multiply the first equation by 5, and the second by 7.)

Summary — How to Get Rid of an Unknown

Here is a summary of the steps in solving a system of two linear equations in two unknowns:

1. Multiply the equations by numbers so that the coefficients of one of the unknown (say x) become equal. Easiest: multiply the first equation by the coefficient of x in the second, and the second equation by the coefficient of x in the first.

2. Subtract one equation from the other (it does not matter which one from which one).
3. We obtain an equation in one unknown — solve it.
4. Put the value of the unknown you found in one of the equations (does not matter which one) to get the value of the yet unfound unknown.

A Second Method: Expressing One Variable in Terms of the Other

Look at the following system:

$$y = x$$

$$x + y = 36$$

This is not really a case of two unknowns: since by the first equation $y = x$, the second equation is nothing but $x + x = 36$, an equation in one variable. The situation is not much more complicated in:

$$y = 3x$$

$$x + y = 36$$

Again, y is redundant. By the first equation it can be expressed in terms of x, and putting this in the second equation we get $x + 3x = 36$, giving $x = 9$. There was really only one unknown, x.

Sometimes, so it turns out, one equation can be used to get rid of one unknown, by expressing it in terms of the other. But is it really "sometimes"? In fact, it is "always". We can always use one equation to express one unknown in terms of the other. This is the idea of a second method for solving a system equations, called "extraction".

Extraction

Solve: $\begin{array}{l} 2x + y = 5 \\ 3x - y = 5 \end{array}$.

Solution: Extract y from the second equation (we could just as well extract it also from the first, or extract x in terms of y from one of the equations). This equation reads $3x - y = 5$, yielding $y = 3x - 5$. We managed to express y in terms of x. Put this in the first equation. (There is no point putting this in the second equation — this we have already used. Can you guess what will happen if we put $y = 3x - 5$ in the second equation? It will not be very useful. Try!) The first equation becomes then: $2x + (3x - 5) = 5$, and collecting terms gives $5x = 10$, namely $x = 2$. Putting this into the equality $y = 3x - 5$ gives $y = 1$.

Solve: $\begin{matrix} x + 4y = -1 \\ 5x - 2y = 17 \end{matrix}$.

Solution: It is best here to extract x from the first equation, since it appears there with coefficient 1. We get: $x = -4y - 1$. Put this in the second equation: $5(-4y - 1) - 2y = 17$, which gives $-22y = 17 + 5 = 22$, namely $y = -1$. Then $x = -4y - 1 = -4 \times (-1) - 1 = 3$.

The extraction can lead to fractions.

Solve: $\begin{matrix} 3x + 4y = -1 \\ 5x - 2y = 7 \end{matrix}$.

Solution: Extracting y from the first equation we get: $4y = -1 - 3x$, or $y = \frac{-1-3x}{4}$. Putting this in the second equation we get

$$5x - 2 \times \frac{-1 - 3x}{4} = 7,$$

which becomes $5x + \frac{1+3x}{2} = 7$, or $6\frac{1}{2}x = 6\frac{1}{2}$, namely $x = 1$. Plugging this into the equality $y = \frac{-1-3x}{4}$ gives $y = \frac{-1-3}{4} = -1$.

Which Method is More Convenient?

There is no universal answer to this question. In systems of more than two equations the extraction method becomes cumbersome, and hence the method of equalizing coefficients is used. For this reason, the computer almost always uses the method of coefficient equalization. The extraction method is particularly convenient if at least one of the coefficients in one of the equations is 1 or (-1). Extracting the unknown which has this coefficient is then easy.

Choose a method convenient for you, and solve: a. $\begin{matrix} 11x - 4y = 4 \\ 6y - 11x = 16 \end{matrix}$
b. $\begin{matrix} 5x - y = 4 \\ 6y - 2x = 1 \end{matrix}$ c. $\begin{matrix} 5x - y = 4 \\ 6x - 2y = 1 \end{matrix}$.

The Number of Solutions of a System of Linear Equations

When solving a system of equations it is most satisfactory if there is one solution, namely one system of values for the unknowns that satisfies the conditions. But this is not always the case. In fact, it is not always so even in equations with one unknown. For example, the equation $0 \times x = 7$ has no solution, since there is no number whose multiplication by 0 will give 7. We usually express it by the fact that $\frac{7}{0}$ is undefined. On the other hand, the equation $0 \times x = 0$ has infinitely many solutions — in fact, every value of x is a solution.

A similar phenomenon happens in systems of two linear equations. But there the situation can be a bit more complicated.

Systems with No Solution

Nobody may suspect that the following system has a solution:

$$x + y = 1$$
$$x + y = 2$$

The same number cannot be both 1 and 2. This system is self contradictory. The contradiction may be more subtle:

$$x + y = 1$$
$$3x + 3y = 6$$

This is the same system as above, with the second equation multiplied by 3. Another system with no solution is $\begin{array}{c} 3x + 5y = 10 \\ 6x + 10y = 21 \end{array}$. Multiplying the second equation by 2 gives $6x + 10y = 20$, so the system is the same as $\begin{array}{c} 6x + 10y = 20 \\ 6x + 10y = 21 \end{array}$ in which the contradiction is evident. The secret was that the left hand side of the second equation in the system $\begin{array}{c} 3x + 5y = 10 \\ 6x + 10y = 21 \end{array}$ is 2 times the left hand side of the first equation, while in the right hand side the proportion is different.

> When the left hand sides are proportionate, and the right hand sides do not relate in the same proportion, there is no solution.

When the Two Equations are Really the Same

When there is the same proportionality all through the equations, left and right sides, it means that these are really the same equations. One of them just repeats the other, so in fact there is only one equation. And one equation with two unknowns has infinitely many solutions.

Example: $\begin{array}{c} x + y = 1 \\ 3x + 3y = 3 \end{array}$. The second equation is obtained from the first upon multiplication by 3. So there is just one equation, $x + y = 1$, having infinitely many solutions like $x = 0, y = 1$; $x = 1, y = 0$; $x = -1, y = 2$.

We have found the following:

> A system of equations in two unknowns may have one solution, no solution, or infinitely many solutions.

The Graphic Point of View

A linear equation like $2x + y = 5$ describes a line in the plane. It is interesting to relate what we saw above to the behavior of lines. Look at the following system of equations:

$$2x + y = 5$$

$$x - y = 1$$

We can draw the two lines in the plane:

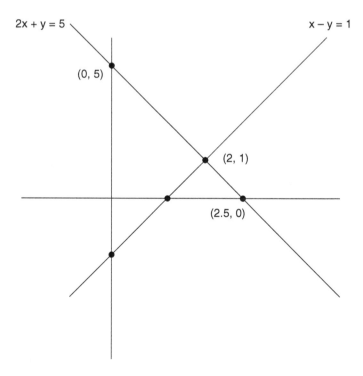

The intersection of the two lines is the solution of the system of equations they represent

The points on the first line satisfy the first equation, and the points on the second line satisfy the second equation. By definition, the meeting point of the lines lies on both lines, meaning that it is a solution to the system. And indeed, the meeting point, $(2, 1)$, satisfies both equations (check).

This provides a third method of solving systems of linear equations — draw the lines, and find the meeting point. It is not a very efficient method, neither is it accurate. But it sheds geometric light on the facts we have just learnt.

The Three Relative Positions of a Pair of Lines

Two lines can be in one of the following three relations to each other:

1. They can meet at a point;
2. They can be parallel;
3. They can be identical.

In the first case, the system of equations has precisely one solution — the meeting point. In the second case there is no meeting point, namely no solution. In the third case, there is really only one line, and every point on this line is a solution for the system.

Solve algebraically and geometrically the system $\begin{array}{l} 2x + y = 1 \\ 6x + 3y = 5 \end{array}$.

Algebraic solution: The two left hand sides are proportionate, the second being 3 times the first, yet the ratio in the right hand side is 5 to 1, so there is no solution.

Geometric solution: From the chapter on the representation of lines we know that the two lines are parallel, since their A, B coefficients (using the notation from that chapter) are proportionate, and their C coefficients do not share the same proportion. So, there is no solution to the system.

Solve geometrically: $\begin{array}{l} x = 3 \\ y = 2 \end{array}$. (This is a self-solving system, I know, but please still draw the two lines!)

Solve geometrically: a. $\begin{array}{l} x = 3 \\ y = x \end{array}$ b. $\begin{array}{l} x + y = 8 \\ x - y = 2 \end{array}$.

Which of the following pairs of lines meet, which are parallel, and which pairs coincide?

a. $\begin{array}{l} x + 2y = 1 \\ x + 2y = 3 \end{array}$ b. $\begin{array}{l} x + 2y = 1 \\ 5x + 10y = 5 \end{array}$ c. $\begin{array}{l} 3x + 2y = 1 \\ 5x + 2y = 3 \end{array}$ d. $\begin{array}{l} 7x - 2y = 1 \\ 14x + 4y = 3 \end{array}$.

A Bit Beyond — Determinants

Linear Algebra

Solving systems of equations played a special role in the history of mathematics. It instigated a whole new area, called "linear algebra". The first to develop it were the Chinese, around 200BC. It reached Europe in the 16-th century, and was intensely studied during the 18-th and 19-th centuries. Nowadays it is one of the most basic mathematical tools.

The Determinant

Look again at the system $\begin{smallmatrix} 3x + 5y = 10 \\ 6x + 10y = 21 \end{smallmatrix}$. We noted that the left hand sides are proportionate. The proportionality is of the coefficients, so that we can really speak of those. Let us put them in a square: $\left(\begin{smallmatrix} 3 & 5 \\ 6 & 10 \end{smallmatrix}\right)$. Such a square is called a (square) matrix. In general, a matrix is a rectangle of numbers. The rectangle can also be with only one column, like $\left(\begin{smallmatrix} 10 \\ 21 \end{smallmatrix}\right)$. A one column matrix is called an (algebraic) "vector" (there are also geometric vectors).

The proportionality of the two rows of the matrix $\left(\begin{smallmatrix} 3 & 5 \\ 6 & 10 \end{smallmatrix}\right)$ means that $\frac{3}{6} = \frac{5}{10}$. Multiplying by the denominators we get $3 \times 10 = 6 \times 5$, or $3 \times 10 - 5 \times 6 = 0$. The expression $3 \times 10 - 5 \times 6$ is called the "determinant" of the matrix $\left(\begin{smallmatrix} 3 & 5 \\ 6 & 10 \end{smallmatrix}\right)$, and is denoted by $\det\left(\begin{smallmatrix} 3 & 5 \\ 6 & 10 \end{smallmatrix}\right)$. In general, the determinant of a 2×2 matrix is the difference between the products of its diagonals. In formula:

$$\det\begin{pmatrix} a & b \\ c & d \end{pmatrix} = a \times d - b \times c.$$

As we have just seen, the determinant checks whether the rows of the matrix are proportionate. If the determinant is 0, then the rows are proportionate.

Using the determinant, check in which matrices the two rows are proportionate: a. $\left(\begin{smallmatrix} 1 & 0 \\ 0 & 1 \end{smallmatrix}\right)$ b. $\left(\begin{smallmatrix} 2 & 0 \\ 0 & 3 \end{smallmatrix}\right)$ c. $\left(\begin{smallmatrix} 2 & 10 \\ 0 & 3 \end{smallmatrix}\right)$ d. $\left(\begin{smallmatrix} 1 & 1 \\ 1 & 1 \end{smallmatrix}\right)$ e. $\left(\begin{smallmatrix} 1 & 10 \\ 10 & 100 \end{smallmatrix}\right)$.

The determinant was known already to the ancient Chinese. Its name was given to it by the French mathematician Cauchy. And indeed, it determines something: whether the system of equations has a single solution or not. If the determinant of the coefficients of a system is 0, it does not have a single solution. It will have either infinitely many solution, or none.

Example: $\det \begin{pmatrix} 2 & 3 \\ 20 & 30 \end{pmatrix} = 2 \times 30 - 3 \times 20 = 60 - 60 = 0$. The system $\begin{smallmatrix} 2x + 3y = 5 \\ 20x + 30y = 50 \end{smallmatrix}$ has infinitely many solutions, while the system $\begin{smallmatrix} 2x + 3y = 5 \\ 20x + 30y = 51 \end{smallmatrix}$ has none.

Crammer's Rule for Solving Systems of Equations

The determinant is a wonderful tool. It can be used to calculate areas and volumes (how — this is beyond the scope of this book). And it can also be used to solve systems of equations, not only determine whether they are solvable. The person who made this discovery, in 1750, was the German mathematician Crammer. I will not explain here why his method works, but I will tell you the way it works. It looks like magic. It may whet your appetite to inquire why. Let me exemplify the method in the following system:

$$2x + 3y = 13$$

$$3x + 7y = 22$$

Before telling you what are Crammer's formulas for the solution, I will tell you what appears in them. You will not be surprised — determinants. One is the determinant of the system, namely $\det \begin{pmatrix} 2 & 3 \\ 3 & 7 \end{pmatrix}$. Another determinant appearing in the formulas is $\det \begin{pmatrix} 13 & 3 \\ 22 & 7 \end{pmatrix}$. Can you guess where did this matrix come from? It is the matrix $\begin{pmatrix} 2 & 3 \\ 3 & 7 \end{pmatrix}$, with the first column replaced by $\begin{pmatrix} 13 \\ 22 \end{pmatrix}$, the right hand side of the system.

Now let me tell you the first of Crammer's formulas. It says that the value of x in the solution to the system is:

$$x = \frac{\det \begin{pmatrix} 13 & 3 \\ 22 & 7 \end{pmatrix}}{\det \begin{pmatrix} 2 & 3 \\ 3 & 7 \end{pmatrix}} = \frac{13 \times 7 - 3 \times 22}{2 \times 7 - 3 \times 3} = \frac{91 - 66}{14 - 9} = \frac{25}{5} = 5.$$

So, the value of x the ratio between the two determinants. In the denominator there is the matrix of coefficients of the system; in the numerator there is the determinant of the matrix in which the first column was replaced.

Before proceeding, try to guess what Crammer's rule for the value of y is. Here it is:

$$y = \frac{\det \begin{pmatrix} 2 & 13 \\ 3 & 22 \end{pmatrix}}{\det \begin{pmatrix} 2 & 3 \\ 3 & 7 \end{pmatrix}} = \frac{2 \times 22 - 3 \times 13}{2 \times 7 - 3 \times 3} = \frac{5}{5} = 1$$

The denominator is the same, but in the numerator there is now the determinant of the matrix in which the coefficients of y are replaced by $\begin{pmatrix} 13 \\ 22 \end{pmatrix}$.

Let me remark that if the determinant of the matrix of coefficients is 0, then in Crammer's rule there appears 0 in the denominator, meaning that there is no single solution.

Let me exemplify Crammer's method in a trivial case, a self solving system:

Example: Use Crammer's formulas to solve $\begin{smallmatrix} x = 8 \\ y = 2 \end{smallmatrix}$.

Solution: Write the system as $\begin{smallmatrix} 1 \cdot x + 0 \cdot y = 8 \\ 0 \cdot x + 1 \cdot y = 2 \end{smallmatrix}$. The matrix of coefficients is $\begin{pmatrix} 1 & 0 \\ 0 & 1 \end{pmatrix}$, whose determinant is 1. By Crammer's formula

$$x = \frac{\det \begin{pmatrix} 8 & 0 \\ 0 & 1 \end{pmatrix}}{\det \begin{pmatrix} 1 & 0 \\ 0 & 1 \end{pmatrix}} = \frac{8 \times 1 - 0 \times 0}{1 \times 1 - 0 \times 0} = 8 \quad \text{and}$$

$$y = \frac{\det \begin{pmatrix} 1 & 0 \\ 0 & 2 \end{pmatrix}}{\det \begin{pmatrix} 1 & 0 \\ 0 & 1 \end{pmatrix}} = \frac{1 \times 2 - 0 \times 0}{1 \times 1 - 0 \times 0} = 2.$$

Of course, you do not need Crammer for this.

Solve: $\begin{smallmatrix} x + y = 8 \\ x - y = 2 \end{smallmatrix}$.

$$\text{Solution: } x = \frac{\det \begin{pmatrix} 8 & 1 \\ 2 & -1 \end{pmatrix}}{\det \begin{pmatrix} 1 & 1 \\ 1 & -1 \end{pmatrix}} = \frac{8 \times (-1) - 1 \times 2}{1 \times (-1) - 1 \times 1} = \frac{-8 - 2}{-1 - 1} = \frac{-10}{-2} = 5,$$

$$y = \frac{\det \begin{pmatrix} 1 & 8 \\ 1 & 2 \end{pmatrix}}{\det \begin{pmatrix} 1 & 1 \\ 1 & -1 \end{pmatrix}} = \frac{1 \times 2 - 1 \times 8}{1 \times (-1) - 1 \times 1} = \frac{2 - 8}{-1 - 1} = \frac{-6}{-2} = 3.$$

What do you get in Crammer's formulas for the system $\begin{smallmatrix} x + y = 8 \\ x + y = 8 \end{smallmatrix}$?

(Answer: $x = y = \frac{0}{0}$, which is meaningless, but it indicates that there are infinitely many solutions.)

Part 13

Quadratic Equations and Quadratic Expressions

Quadratic Equations

The first type of equations we studied was linear equations. One reason is that they appear frequently in real life problems. Another reason is that like the proverbial drunkard who looked for his lost key under the light pole "because here there is light" we chose to study equations that are relatively easy to solve. Moving a bit away from the light (but not much) we get to quadratic equations, namely equations in which the unknown may be squared.

An equation is called "quadratic" if the unknown appears in it at most to the power 2. Here are some examples of quadratic equations:

$$x^2 = 9$$
$$x^2 + x = 2$$
$$3x^2 - x = 10$$
$$x^2 + 2x + 1 = 0$$

Each of these equations has a solution which is a small integral number. Try to guess the solutions.

A Quadratic Equation May Have Two Solutions

A linear equation has a single solution, or no solution at all, or infinitely many solutions. A quadratic equation, by contrast, may have two solutions. For example, $x^2 = 9$ has the two solutions $x = 3$ and $x = -3$. Here are some more examples. In each of them try to guess the two solutions. The guessing game will give you a good feel for the nature of quadratic equations. Hint: the solutions are small integers. Try to find links between equations, and see whether they are reflected also in the solutions.

Solve: a. $x^2 - 3x + 2 = 0$ b. $x^2 + 3x + 2 = 0$ c. $x^2 - 4x + 3 = 0$ d. $x^2 + 4x + 3 = 0$ e. $x^2 - 2x - 3 = 0$ f. $x^2 + 2x - 3 = 0$ g. $x^2 + x + 1 = 3$ h. $x^2 - x + 1 = 3$.

(Solutions: a. 1 and 2, b. (-1) and (-2).)

189

A Quadratic Equation May Have No Solution

The square of any number, be it positive negative or zero, is always non-negative. So, whatever the value of x is, the expression $x^2 + 1$ is positive. It cannot be 0. So, the equation $x^2 + 1 = 0$ has no solution. Moving the 1 to the right hand side, the equation is $x^2 = -1$, so what we have just said is: (-1) has no square root (among the numbers we deal with, the "real numbers") — a fact we have already encountered.

It may be viewed as a sad fact of life, that there are quadratic equations with no root. But it is not really sad. It makes for interesting life.

Quadratic Equations with Just One Root

Finally, there are equations with just one root. For example, $x^2 = 0$, whose only solution is 0. But also $(x - 5)^2 = 0$ has just one root: $x - 5$ must be 0, so $x = 5$ is the only solution!

If we open parentheses, the equation $(x-5)^2 = 0$ becomes $x^2 - 10x + 25 = 0$, where it is not so easy to see that there is just one solution. How to see that an equation has two roots, one root or none? Wait a bit. We shall get there.

Roots and How to Write Them

The solution of the quadratic equation $x^2 = 2$ is $x = \sqrt{2}$, namely the root of 2. This gave all solutions of quadratic equations the name "roots". "Root", in this context, is just a synonym of "solution". So, for example, the roots of $x^2 - 3x + 2 = 0$ are $x = 1$ and $x = 2$. We write this as follows: $x_1 = 1$ and $x_2 = 2$ (the order does not matter). Or, sometimes even shorter: $x_{1,2} = 1, 2$.

Let us now return to the equation $x^2 = 2$. In fact, it has two solutions: $\sqrt{2}$ and $-\sqrt{2}$. (Remember that $\sqrt{2}$ is positive. It is defined as the positive number whose square is 2.) This we write as follows: $x_{1,2} = \pm\sqrt{2}$.

Write this way the roots of $(x - 1)(x + 1) = 0$ and of $x^2 - 1 = 0$.
Write this way the two roots of a. $x^2 = 10$ b. $x^2 = 9$ c. $(x+3)(x-3) = 0$ (you need the root sign only in the first).

Inventing Quadratic Equations

At this point experimentation is a must. The best way to understand equations is to invent them (the second is to guess solutions and try plugging your guesses in, to check whether they work). In class, I ask what is the simplest quadratic equation. After some discussion we get to $x^2 = 0$, whose only solution is $x = 0$. Next in simplicity is $x^2 = 1$. It has two solutions: $x = 1$ and $x = (-1)$. As a next simple example the students suggest $x^2 = 2$. We struggle a bit, to find that there is no simple solution. Those who know roots say that it is $\sqrt{2}$.

Can you invent an equation whose solution is $x = 10$? Of course — $x^2 = 100$.

Let us now turn to more complicated equations. What will you put in the right hand side of the equation $x^2 + x = \cdots$ so that $x = 3$ is a solution? Most students find this a non-trivial task, until they understand — the number on the right should be $3^2 + 3$, namely 12, so the equation is $x^2 + x = 12$. Does this equation have another solution? Try negative numbers. (Yes, it does: $x = -4$.)

Experimentation lessons like this are best run with a full class, not individually. Class discussion is an indispensable tool for reaching insights.

Inventing Equations by Two Solutions

Why can a quadratic equation have two solutions?

Look at the following equation: $(x - 2)(x - 3) = 0$. The product of two numbers is 0 if one of them is 0. So, for x to be a solution, either $x - 2 = 0$ or $x - 3 = 0$. In the first case $x = 2$ and in the second case $x = 3$. So, this equation has two solutions.

This is nice, but what does it have to do with quadratic equations? To see this, open the parentheses on the left side.

$$(x - 2)(x - 3) = x \times x - 3 \times x - 2 \times x - 2 \times (-3) = x^2 - 5x + 6.$$

So, our equation is $x^2 - 5x + 6 = 0$. A quadratic equation! Not surprising, because the left side was a product of two terms.

Solve: $(5x - 5)(14x - 7) = 0$.

Solution: The product is 0 if either $5x - 5 = 0$, which happens if $x = 1$, or $14x - 7 = 0$, which happens when $x = \frac{1}{2}$.

At this point we know how to find an equation by its two solutions.

Find an equation whose two solutions are $x = 5$ and $x = 1$.

Solution: $(x - 5)(x - 1) = 0$, and opening parentheses: $x^2 - 5x - 1x + 5 = 0$, namely

$$x^2 - 6x + 5 = 0.$$

It is instructive to see what happens if we reverse signs:

Find an equation whose two solutions are $x = -5$ and $x = -1$. Solution: $(x+5)(x+1) = 0$, and opening parentheses: $x^2 + 5x + 1x + 5 = 0$, namely

$$x^2 + 6x + 5 = 0.$$

Just the coefficient of x changed its sign!

Can you find another equation with solutions $x = -5$ and $x = -1$?

Answer: multiply both sides of the equation by (say) 3. You get: $3x^2 + 18x + 15 = 0$.

Multiplying the two sides of an equation by the same number does not change the solutions.

In class I now devote a lot of time to inventing equations by their solutions, and to investigating the properties of the equations. How does an equation look if one of its solutions is $x = 0$? How do you produce a quadratic equation with only one solution?

What are the solutions of $x^2 + x = 0$?

(Hint: write the equation as $x(x+1) = 0$.)

Find an equation whose solutions are $x = 0$ and $x = -1$.
Find a quadratic equation whose only solution is $x = 10$.

Solution: The left hand side is a product, in which one of the terms is $x - 10$. How can it be that the other term does not give another solution? Of course — the other term should also be $x - 10$. The equation is $(x - 10)(x - 10) = 0$, namely $(x - 10)^2 = 0$.

Find an equation whose solutions are $x = 10$ and $x = 20$, and in which the coefficient of x^2 is 3.

(Answer: $3(x - 10)(x - 20) = 0$, namely $3x^2 - 90x + 600 = 0$.)

Write a quadratic equation whose only solution is $x = -5$, and in which the free term is 10.

Canonical Form

The previous section taught us that it is a good idea to have 0 on the right hand side of the quadratic equation. This is indeed customary. This form, in which there is 0 on the right, is called "canonical" (meaning "standard"). Of course, every equation can be written canonically:

Transform $x^2 + x + 1 = 3$ to canonical form.

Answer: $x^2 + x - 2 = 0$.

Write the following equations in canonical form: $x^2 + x - 2 = 3x^2 - 2x + 5$.

Answer: $x^2 + x - 2 - (3x^2 - 2x + 5) = 0$, which is: $-2x^2 + 3x - 7 = 0$.

The general canonical form of a quadratic equation is:

$$ax^2 + bx + c = 0.$$

Often these very letters are used for the coefficients — a for the coefficient of x^2, b for the coefficient of x and c for the free term.

Monic Form

We saw that there are many equations in canonical form with the same solutions. Of those, we want to choose a particularly convenient one, in which the coefficient of x^2 is 1. It is called then "monic". "Mono" is one, as in "Monogamy".

Transform $3x^2 + x - 6 = 0$ to monic form.

Solution: Divide both sides by 3, to get $x^2 + \frac{1}{3}x - 2 = 0$.

Write $5x^2 - 15x + 10 = 0$ and $5x^2 - 10x + 1 = 0$ in monic form. Write a, b and c in each case.

The Vieta Formulas

Look at the equation $(x - 5)(x - 7) = 0$. Its roots are 5 and 7. Opening the brackets it is: $x^2 - 5x - 7x + 5 \times 7 = 0$, namely $x^2 - 12x + 35 = 0$. You have probably followed the way the coefficients (-12) and 35 were formed. The coefficient of x, (-12), is the sum of the two roots, with a minus sign; the free term, 35, is their product.

These two observations are named after the Italian mathematician Francisco Vieta (1490–1550. His name is often written the French way, Viete). If you live in the right period, you may get your name on a mathematical discovery that is not necessarily hard! Let us write it formally:

Theorem: if the roots of the monic equation $x^2 + px + q = 0$ are x_1 and x_2, then $x_1 + x_2 = -p$ and $x_1 x_2 = q$.

Now, when asked to find an equation by its roots, we have an easier life:

Find an equation whose roots are 3 and 5. By the Vieta formulas, the equation $x^2 - 8x + 15 = 0$ fulfils the requirement. (It is not the only

equation to do so, for example $2x^2 - 16x + 30 = 0$ does this as well —
why?)

It is sometimes possible to use the Vieta formulas to guess the solutions.

Solve $x^2 - 10x + 21 = 0$.

Solution: We are looking for two numbers whose sum is 10, and their
product is 21. Not hard to guess (the product gives the better clue), these
are 7 and 3, which are therefore the roots.

Solve $x^2 - 7x + 10 = 0$.

We have to find numbers whose product is 10 and their sum 7. These
are 5 and 2.

Solve $x^2 + 7x + 10 = 0$.

The product is again 10, and the sum is (-7). The appropriate numbers
are -5 and -2.

Use the Vieta formulas to guess solutions for the following equations:

> a. $x^2 + 7x + 6 = 0$ b. $x^2 - 7x + 6 = 0$ c. $x^2 + 8x + 12 = 0$
> d. $x^2 + 15x + 26 = 0$ e. $x^2 - 15x + 26 = 0$.

The Signs of the Roots

Look at the equation $x^2 - 13x + 30 = 0$. The product of the roots is 30, which
is positive. The product of two numbers is positive if they have the same
sign, namely both are positive or both are negative. Which of the two cases is
it? Since the sum of the roots is 13, which is positive, they are both positive.
It is not hard to guess: these are 10 and 3. In the equation $x^2 + 13x + 30$, on
the other hand, the product of the roots is 30, so they are of the same sign,
and their sum is (-13), so they are both negative. They are (-10) and (-3).

In general, in a monic equation $x^2 + px + q = 0$, if the free term q is
positive, the two roots have the same sign, and this sign is opposite to that
of p. If q is negative, the two roots are of opposite signs.

> Solve using the Vieta formulas: a. $x^2 - 16x + 39 = 0$ b. $x^2 - 3x - 10 = 0$
> c. $x^2 + 3x - 10 = 0$.

The Vieta Formulas for General Equations

If you wish to use the Vieta formulas and the equation is not monic, just
normalize. Namely, divide the equation by the coefficient of x^2.

Example: what is the sum of the roots in the equation $2x^2 + 5x + 1 = 0$, and what is their product?

Solution: Normalize, by dividing both sides of the equation by 2. The equation becomes $x^2 + \frac{5}{2}x + \frac{1}{2} = 0$. The sum of the roots is $-\frac{5}{2}$ (the coefficient of x, in minus sign) and the product is $\frac{1}{2}$.

In general, the equation $ax^2 + bx + c = 0$, when normalized, becomes

$$x^2 + \frac{b}{a}x + \frac{c}{a} = 0.$$

The sum of the roots is therefore $-\frac{b}{a}$, and their product is $\frac{c}{a}$.

Guess one of the roots of the equation $5x^2 - 6x + 1 = 0$, and use the first root to find the second.

Solution: It is not hard to see that $x = 1$ is a solution. By the Vieta formula for the sum of roots, the sum of the roots is $\left(-\frac{6}{5}\right)$, so the other root must be $\left(-\frac{6}{5}\right) - 1 = -2\frac{1}{5}$.

Solving Systematically

The Vieta formulas are nice for guessing. But we need a systematic solution, and this we do now. We shall learn a formula for the roots — a formula known already to the Babylonians, around 1000BC ! I could write you the formula right away, and then check that it works, but I prefer to lead you to it step by step.

Equations with Just a Square

You know how to solve an equation of the form $x^2 = 9$. You just take a root, and find $x_{1,2} = \pm 3$. But then, it is not hard to solve also:

$$(x + 1)^2 = 9$$

Taking roots, we get: $x + 1 = \pm 3$. This means that either $x + 1 = 3$, namely $x = 3 - 1 = 2$, or $x + 1 = -3$, meaning $x = -3 - 1 = -4$. So, the roots are $x_1 = 2, x_2 = -4$. (Let us check the first root: $(x_1 + 1)^2 = (2 + 1)^2 = 3^2 = 9$. Check also the second root, (-4).)

Solve: $(x - 2)^2 = 100$.

Solution: $x - 2$ is either 10, in which case $x = 10 + 2 = 12$; or $x - 2$ is (-10), in which case $x = -10 + 2 = -8$. The roots are $x_1 = 12, x_2 = -8$.

Solve: a. $(x+10)^2 = 1$ b. $(x+10)^2 = 0$ c. $(x-5)^2 = 0$ d. $(x+10)^2 = -1$ e. $(x-5)^2 = 25$ f. $(x+8)^2 = 0$.

Note — one of these equations has no root!

When the right hand side is not a square of an integer, we have to use the root notation:

Solve: $(x+5)^2 = 7$.

Solution: Taking roots, we get $x + 5 = \sqrt{7}$ or $x + 5 = -\sqrt{7}$. In the first case $x = \sqrt{7} - 5$, and in the second $x = -\sqrt{7} - 5$. We write this concisely: $x_{1,2} = -5 \pm \sqrt{7}$.

Solve: a. $(x-5)^2 = 5$ b. $(x-5)^2 = 2$ c. $(x-5)^2 = 10$ d. $(x-2)^2 = 5$.

In general, the roots of $(x + a)^2 = b$, when $b \geq 0$ (that is, when b is non-negative), are $x = \pm\sqrt{b} - a$. When b is negative, there is no root.

Use this formula to solve $(x + 10)^2 = 12$.

Completing to a Square

The trick for solving a general quadratic equation is to bring it to the form $(x + a)^2 = b$. This is called "completing to a square", the square being $(x + a)^2$.

Solve $x^2 + 6x + 5 = 0$.

Solution: By the identity $(a+b)^2 = a^2 + 2ab + b^2$, we know that $(x+3)^2 = x^2 + 6x + 9$. So, $x^2 + 6x$, which appears in the equation, is almost $(x + 3)^2$. In fact, $x^2 + 6x = (x + 3)^2 - 9$. So,

$$x^2 + 6x + 5 = (x + 3)^2 - 9 + 5 = (x + 3)^2 - 4,$$

the equation can be written as $(x+3)^2 - 4 = 0$, namely $(x+3)^2 = 4$, which is just the form we wanted. The solutions are $x + 3 = \pm 2$, namely $x = \pm 2 - 3$, meaning $x_{1,2} = -2, -5$.

Where did the $(x + 3)^2$ appear from? The hint was in $x^2 + 6x$. The "3" was just half of 6, the coefficient of x. If the equation is normalized, namely it is $x^2 + px + q = 0$, we compare it with $\left(x + \frac{p}{2}\right)^2$. This is the square hiding behind the equation.

An interim summary:

1. For the time being we look at a normalized equation, $x^2 + px + q = 0$.
2. We "complete the square $\left(x + \frac{p}{2}\right)^2$", meaning that we write $x^2 + px + q$ using $\left(x + \frac{p}{2}\right)^2$.

Since this is important, I want to give two more examples before going to the general solution of a quadratic equation.

Solve $x^2 + 14x + 24 = 0$.

Solution: Here $p = 14$ and $\frac{p}{2} = 7$. Since $x^2 + 14x = (x + 7)^2 - 49$, we have $x^2 + 14x + 24 = (x + 7)^2 - 49 + 24 = (x + 7)^2 - 25$. So, the equation is $(x + 7)^2 - 25 = 0$, meaning that $(x + 7)^2 = 25$, whose roots are $x_{1,2} = \pm 5 - 7$, hence $x_{1,2} = -2, -12$.

Solve: $x^2 - 20x + 64 = 0$.

Solution: Here $p = -20$, $\frac{p}{2} = -10$. Write $x^2 - 20x + 64 = (x - 10)^2 - 100 + 64 = (x - 10)^2 - 36$. So the equation is $(x - 10)^2 - 36 = 0$, namely $x - 10 = \pm 6$, hence $x = 10 \pm 6$.

Solve: a. $x^2 - 20x + 100 = 0$ b. $x^2 + 8x + 16 = 0$ c. $x^2 + 8x - 9 = 0$
d. $x^2 + 8x + 17 = 0$ e. $x^2 - 16x + 64 = 25$ f. $x^2 + 3x + \frac{9}{4} = 5$
g. $x^2 + x + \frac{1}{4} = 9$.

The General Formula for Normalized Equations

Having done enough examples, we are ready to generalize: we are going to find a formula for the solution of any quadratic equation. We start from a normalized equation:

$$x^2 + px + q = 0.$$

Trying to replace $x^2 + px$ by $\left(x + \frac{p}{2}\right)^2$, we find that $x^2 + px = \left(x + \frac{p}{2}\right)^2 - \frac{p^2}{4}$ (the last term cancelling out the redundant term in $(x + \frac{p}{2})^2$). So, the equation is

$$\left(x + \frac{p}{2}\right)^2 - \frac{p^2}{4} + q = 0,$$

which, upon moving sides, becomes

$$\left(x + \frac{p}{2}\right)^2 = \frac{p^2}{4} - q,$$

which is the familiar "complete square" form. We only have to take roots:

$$x + \frac{p}{2} = \pm\sqrt{\frac{p^2}{4} - q},$$

and one more side-moving,

$$x_{1,2} = -\frac{p}{2} \pm \sqrt{\frac{p^2}{4} - q}.$$

That's it, this is the solution. But there is another twist, not a must, but customary. We take a common denominator 4 inside the root, and then use the fact that $\sqrt{4} = 2$. Here:

$$x_{1,2} = -\frac{p}{2} \pm \sqrt{\frac{p^2}{4} - q} = -\frac{p}{2} \pm \sqrt{\frac{p^2 - 4q}{4}}$$

$$= -\frac{p}{2} \pm \frac{\sqrt{p^2 - 4q}}{2} = \frac{-p \pm \sqrt{p^2 - 4q}}{2}.$$

Summarizing:

$$x_{1,2} = \frac{-p \pm \sqrt{p^2 - 4q}}{2}.$$

And this is really it.

Solve $x^2 + 6x + 5 = 0$. Solution:

$$x_{1,2} = \frac{-p \pm \sqrt{p^2 - 4q}}{2} = \frac{-6 \pm \sqrt{6^2 - 4 \times 5}}{2}$$

$$= \frac{-6 \pm \sqrt{36 - 20}}{2} = \frac{-6 \pm 4}{2} = -5, -1.$$

You may remember that we solved this equation before, but this time we did it automatically, without having to think.

Use the formula to solve a. $x^2 + 2x = 2$ b. $x^2 + 2x = 35$ c. $x^2 + 2x = 0$ d. $x^2 + 2x = 9999$.

The Formula for General Equations

To solve a general equation of the form $ax^2 + bx + c = 0$, we normalize it: dividing by a we get $x^2 + \frac{b}{a}x + \frac{c}{a} = 0$. Now we can use the formula for the roots of a normalized equation, with $p = \frac{b}{a}$ and $q = \frac{c}{a}$. The formula gives

$$x_{1,2} = \frac{-\frac{b}{a} \pm \sqrt{\frac{b^2}{a^2} - 4\frac{c}{a}}}{2},$$

and taking a^2 as a common denominator in the root, we get

$$x_{1,2} = \frac{-\frac{b}{a} \pm \sqrt{\frac{b^2 - 4ac}{a^2}}}{2} = \frac{-\frac{b}{a} \pm \frac{\sqrt{b^2 - 4ac}}{a}}{2} = \frac{-b \pm \sqrt{b^2 - 4ac}}{2a}.$$

Summarizing:

$$x_{1,2} = \frac{-b \pm \sqrt{b^2 - 4ac}}{2a}.$$

Examples: 1. $x^2 = 0$. In this case, $a = 1$, $b = 0$, $c = 0$, so the formula gives $x_{1,2} = \frac{-0 \pm \sqrt{0}}{2} = 0$, which of course we knew.

2. $x^2 = 9$. Here $a = 1$, $b = 0$, $c = -9$, so the formula gives

$$x_{1,2} = \frac{-0 \pm \sqrt{0^2 + 4 \times 1 \times 9}}{2} = \frac{\pm\sqrt{36}}{2} = \frac{\pm 6}{2} = \pm 3.$$

In both examples you see that the formula is not cheating.

Use the formula to solve $4x^2 + 5x - 9 = 0$.
Use the formula for the roots to prove the Vieta formulas. (The sum formula is easy, can you do it without using pen and paper?)

The Discriminant

Under the root sign in the formula there appears the expression $b^2 - 4ac$. It is called the "discriminant" of the equation, and is denoted by the Greek letter for "D", which is Δ. It got the name from the fact that it discriminates equations, in the sense that it tells solvable equations from unsolvable ones. It does so very simply:

A. If Δ is positive, there are two solutions, one with $+\sqrt{\Delta}$, and the other with $-\sqrt{\Delta}$.
B. If $\Delta = 0$ then $\sqrt{\Delta} = 0$, so adding it is the same as subtracting it. So, the equation has just one square root solution.
C. If Δ is negative, it has no root, so the equation has no solution.

Examples: 1. $x^2 = 9$. Here $a = 1$, $b = 0$, $c = -9$, hence $\Delta = b^2 - 4ac = 0^2 - 4 \times 1 \times (-9) = 0 - (-36) = 36$, which is positive, so there are two solutions. (Of course there are — $+3$ and -3.)

2. $(x - 5)^2 = 0$. We know that there is just one root, 5, so the discriminant should be 0. Indeed, the equation is $x^2 - 10x + 25 = 0$, so $a = 1$, $b = -10$, $c = 25$, and $\Delta = (-10)^2 - 4 \times 1 \times 25 = 100 - 100 = 0$.

Calculate the discriminant in $4x^2 + 4x + 1 = 0$ and of $4x^2 + 4x - 1 = 0$, and determine how many roots each equation has.
In each of the following equations there appears a number m that can vary. In each equation determine for which values of m there are two roots to the equation, and for which value of m there is precisely

one solution:

a. $x^2 + m = 0$ b. $x^2 + 2x + m = 0$ c. $x^2 - 2x + m = 0$ d. $x^2 + 3x + m = 0$
e. $x^2 + 10x + m = 0$ f. $x^2 - 10x + m = 0$.

When a and c Have Opposite Signs

Look at the equation $10x^2 + 10x - 999$. We have $a = 10$ and $c = -999$, so $\Delta = 10^2 - 4 \times 10 \times (-999) = 10^2 + 4 \times 10 \times 999$, which is the sum of two positive numbers, so it is positive, so the equation has two roots. What is happening here is that a and c have different signs, so $(-4ac)$ is positive. But b^2, like every square, is non-negative. So, the discriminant, namely $b^2 - 4ac$, is positive.

The Same Conclusion, Without Discriminants

There is another way of seeing this, not necessitating the discriminant. Look at the expression $10x^2 + 10x - 999$. For $x = 0$ we get (-999), which is negative. If we put very large x, say $x = 1000$, the significant term is $10x^2$ (for $x = 1000$ this is 10 million), which makes the whole expression $10x^2 + 10x - 999$ positive (subtracting 999 from 10,000,000 will not give a negative number). So, going from $x = 0$ to $x = 1000$ takes $10x^2 + 10x - 999$ from negative to positive. Somewhere on the way it must be 0, just as when you go from the Dead Sea to Jerusalem you must go through a place which is precisely at sea level. This means that there is some value of x for which $10x^2 + 10x - 999 = 0$. Namely, the equation has a solution.

Quadratic Functions

Just as linear functions are interesting in their own right, not only in connection with linear equations, so quadratic functions are interesting in their own right. They often appear in real life.

A quadratic function is given by a quadratic expression, like $f(x) = x^2$, or $k(x) = 3x^2 + 2x - 5$. There are many interesting questions to ask about such functions: how do their graphs look like? When is the function ascending (increasing) and when is it descending (decreasing)? When is it the smallest, and when is it the largest? All this is closely connected to quadratic equations.

The Function $f(x) = x^2$

The simplest quadratic function is $f(x) = x^2$. Here is its graph:

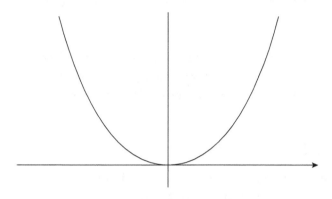

The graph of $f(x) = x^2$

The shape of this graph is called "parabola", a word taken from Greek and meaning "parable" (so, the name has no connection to the shape).

Here are some properties of the function, and their manifestation on the graph:

a. It is non-negative, namely $f(x) \geq 0$ for all x. (The graph does not go below the x axis.)

b. For every positive number m there are two values of x for which $f(x) = m$, namely $x = \sqrt{m}$ and $x = -\sqrt{m}$.

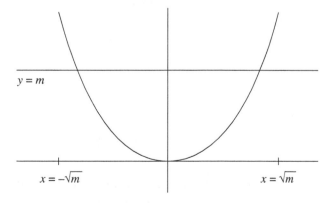

Drawing, with line parallel to the x axis at height m

c. $f(0) = 0$, meaning that the graph goes through the origin. Also, this is the only intersection of the graph with the x axis.

d. The function (its graph) is symmetric with respect to the y axis, meaning that $f(-x) = f(x)$. A function satisfying this property is called an "even function".

Can you guess why even functions are called so? Hint: which of the following functions are even: $f(x) = 5$, $g(x) = x$, $k(x) = x^2$, $h(x) = x^3$, $j(x) = x^4$, $i(x) = x^5$? Give another example of an even function.

Lifting and Dropping

The graph of the function $g(x) = x^2 + 3$ is obtained from the graph of $f(x) = x^2$ by shifting up 3 units, so its graph is like this:

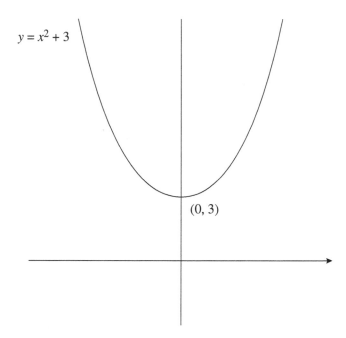

The graph of $g(x) = x^2 + 3$

It does not meet the x axis, which means that $x^2 + 3$ is never 0, in other words, the equation $x^2 + 3 = 0$ has no solution, which of course we know.

How about $k(x) = x^2 - 3$? The graph is shifted 3 units down, and it does meet the x axis in two points, $x = \pm\sqrt{3}$.

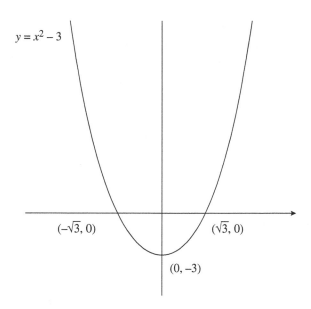

Drawing of $k(x) = x^2 - 3$

Shifting Sideways

Look at the function $g(x) = (x - 3)^2$. Like $f(x) = x^2$, it is always non-negative, and it also meets the x axis at one point: $x = 3$. Its graph looks like this:

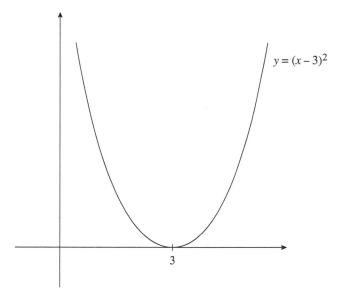

Graph of $g(x) = (x - 3)^2$

So, the graph is just like that of $f(x) = x^2$, shifted 3 units to the right. This is no surprise. For example, $g(5) = (5 - 3)^2 = 2^2 = f(2)$, so at $x = 5$ the graph of g looks like the graph of f at $x = 2$. The graph of g looks 3 units to the left, sees what the graph of f does there, and imitates.

Similarly, the graph of $k(x) = (x + 2)^2$ is like the graph of $f(x) = x^2$, shifted 2 units to the left. It meets the x axis at $x = -2$.

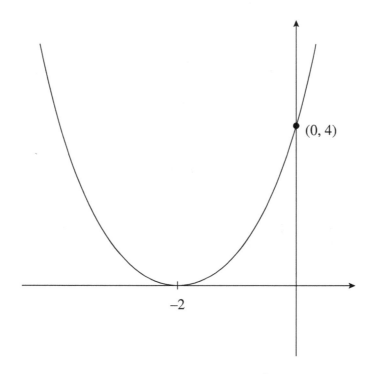

The graph of $k(x) = (x + 2)^2$

We can do both lifting and shifting. For example, the graph of $h(x) = (x + 3)^2 - 4$ is obtained from the graph of $f(x) = x^2$ by shifting 3 units left, and dropping 4 units.

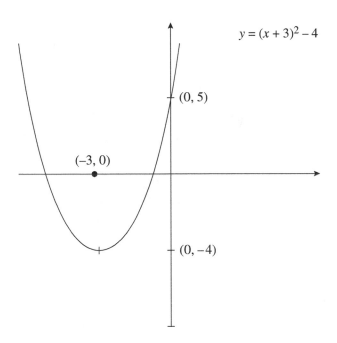

$$y = (x+3)^2 - 4$$

$(0, 5)$

$(-3, 0)$

$(0, -4)$

Graph of $h(x) = (x+3)^2 - 4$

You Can Get Every Parabola by...

Let $h(x) = x^2 + 6x + 16$. We can write it as:

$$x^2 + 6x + 16 = (x+3)^2 - 9 + 16 = (x+3)^2 + 7,$$

which is just shifting $f(x) = x^2$ left 3 units, and lifting it 7 units.

> Every normalized quadratic function can be obtained from $f(x) = x^2$ by shifting and lifting.

Waving Wings

And finally, to non-normalized quadratic functions. For example, $g(x) = 2x^2$. How does its graph look like? Here is the graph, together with the graph of $k(x) = x^2$ and $y = \frac{x^2}{2}$.

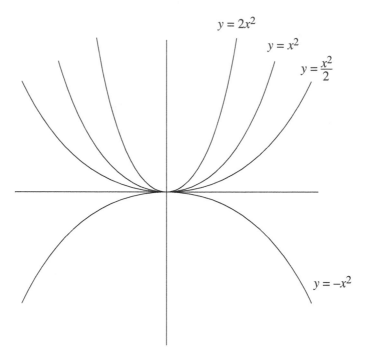

The graph of $y = x^2$ together with friends

It is as if the parabola $y = x^2$ waves its wings up. A general parabola is obtained from a normalized parabola by multiplying by a constant. And this makes the parabola "wave its wings". Wave up, if it is multiplied by a number larger than 1, and down if the number is smaller than 1.

Smiling Parabolas and Sad Parabolas

An interesting thing happens when the coefficient of x is negative. The last drawing also includes the graph of $y = -x^2$. The parabola waves its wings down. It is "sad", as opposed to the "smiling" parabola $y = x^2$.

Why do we drop the corners of our mouths when we are sad?

Why do we raise the corners of the mouth when we are happy, nobody really knows. But why we lower them when sad, we do know. This was ingeniously explained by Charles Darwin, in his 1873 book "The expression of emotions in man and animals". It all starts in wishing to cry. When crying, the mouth is opened by muscles pulling the corners of the mouth both up and down. When we are just sad, not quite crying, we control the

---

tendency to open the mouth. But there is less control over the muscles lowering the corners, and so only they operate, pulling the corners down.

The following two pictures demonstrate this beautifully. They show a 10 months old baby starting to cry. On the left, he is still controlling his wish to cry, so only the lowering of the corners of the mouth occurs. On the right, there is already full blown crying, and the corners are pulled both up and down.

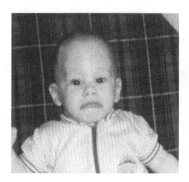 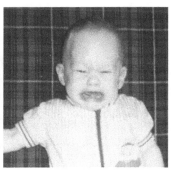

The "happiness" and "sadness" of a parabola are determined by the coefficient of x^2. For example, the graph $y = 10x^2 - 1000x + 10$ is "smiling", because for very large x the expression $10x^2 - 1000x + 10$ is positive, since the term $10x^2$ is the dominant one. So, going far to the right and to the left, the parabola goes up. A similar argument shows that the "wings" of $y = -2x^2 + 1000x$ must point down (or just argue that the wings of $y = 2x^2 - 1000x$ point up, so the wings of its negative, $y = -2x^2 + 1000x$, point down).

The Apex of a Parabola

The lowest point on a smiling parabola, or the highest point on a sad parabola, is called the "apex" of the parabola. For example, the apex of $y = x^2$ is at $(0, 0)$. The apex of $y = x^2 + 5$ is at $(0, 5)$.

How do you find the apex? Of course, it is enough to find the x value of the apex. In the drawing below there is a parabola meeting the x axis. As usual, let us write the parabola in the form $y = x^2 + bx + c$. The two meeting points with the x axis are the points at which the value of y is 0, namely they are the roots of the equation $ax^2 + bx + c = 0$. As can be seen from the drawing, the x value of the apex is in the middle between them, namely the sum of the roots divided by 2. Now, by Vieta's formula, the sum of the roots is $\frac{-b}{a}$. So, the apex is at:

$$x = \frac{-b}{2a}.$$

If the parabola does not meet the x axis, just shift it up or down until it does. This does not change the x value of the apex, neither does it change a or b, so the formula remains valid.

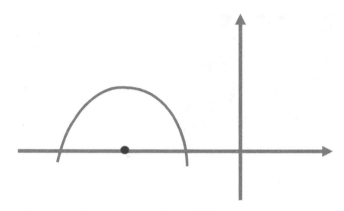

The apex of a parabola is in the middle between its two roots
(if there are no roots, move the parabola up or down until there are)

Where is the apex of $g(x) = x^2 + 5$?

Answer: here b, the coefficient of x, is 0, so by the formula the apex is at 0.

Indeed, this is just parabola $y = x^2$ shifted up, and shifting up does not change the x value of the apex.

Where did we meet the expression $\frac{-b}{2a}$?

(Hint: in the solution of a quadratic equation. Can you explain why this number is in the middle between the two roots of the quadratic expression?)

When Does a Parabola Meet the x Axis?

The parabola $y = x^2 - 1$ meets the x axis at $(-1, 0)$ and $(1, 0)$. The parabola $y = x^2 + 5$, by contrast, lies entirely above the x axis and the parabola $y = -x^2 - 5$ entirely below it. Which parabolas meet the x axis, and which don't? The answer is simple: a parabola $y = ax^2 + bx + c$ meets the x axis at a point x for which $ax^2 + bx + c = 0$. So, there are such meeting points if and only if there is a solution to the quadratic equation. And this happens if and only if $\Delta \geq 0$. If $\Delta = 0$, there is one solution, which means that the parabola is tangent to the x axis.

For each of the following parabolas find, without drawing it, whether it meets the x axis or not; and if it does, in how many points:
a. $y = -x^2 - 2x + 4$ b. $y = -x^2 + 2x - 4$ c. $y = (x - 2)(x - 3)$
d. $y = (x - 1)^2$.

Part 14

Inequalities

Sets of Solutions

Let us recall that "$x < 3$" means "x is smaller than 3", and "$x \leq 3$" means "x is smaller than or equal to 3".

Compare now the equation $x + 1 = 3$ with the inequality $x + 1 < 3$. The first has just one solution, $x = 2$. The second has many solutions: $x = 0$, -1, -2.5 and so on — infinitely many solutions. So, what do we mean by "solving" an inequality like $x + 1 < 3$? It means finding the set of x's that satisfy this inequality. And "finding the set" means describing it succinctly. For example, writing "$x < 2$" instead of "$x + 1 < 3$".

Drawing Sets of Numbers

I want to introduce some notation concerning sets of numbers.

Example: $\{x : x < 1\}$ denotes the set of numbers smaller than 1.

The ":" should be read as "satisfying", so this notation says: "the set of all numbers x satisfying the condition $x < 1$".

Writing $3 < x < 5$ means: "$x < 5$ and $x > 3$". We use one expression to write two conditions at once.

Draw the set $\{x : -3 < x < 5\}$.

Strong and Weak Inequatlities

There is distinction between "$<$", meaning "strictly smaller than", and "$\leq$", meaning "smaller than or equal to". The additional line at the bottom is borrowed from the equality sign.

Allowing equality makes the inequality "weak". When it is strict, we say that it is "strong".

Find a number a satisfying $a \geq 3$ but not $a > 3$.
Is there a number x satisfying both $x \leq 2$ and $x \geq 2$?
Is there a number x satisfying both $x \leq 2$ and $x > 2$?

Drawing Strong and Weak Inequalities

The distinction between strong and weak inequalities should be reflected also in the drawing of the sets. This is done by the use of (round) parentheses to signify "the endpoint does NOT belong to the set", and [square] brackets to signify "the endpoint belongs". For example:

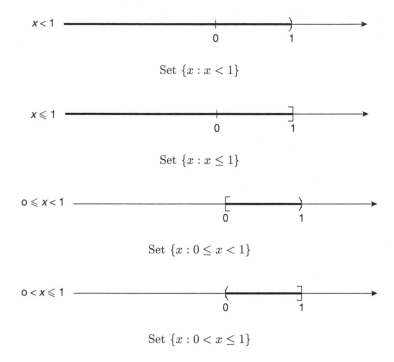

Set $\{x : x < 1\}$

Set $\{x : x \leq 1\}$

Set $\{x : 0 \leq x < 1\}$

Set $\{x : 0 < x \leq 1\}$

Draw the sets $\{x : 0 \leq x \leq 1\}$ and $\{x : 0 < x < 1\}$

Notice the direction of the parentheses or brackets: it is in the direction of the set, indicating which side the set is.

Intervals

A part of the real line enclosed between two points (numbers) is called an "interval". Parentheses and brackets make for easy notation of intervals. For example, $[3, 5)$ is the interval $\{x : 3 \leq x < 5\}$. We say that this interval is closed on the left, and open on the right.

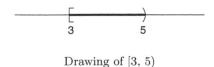

Drawing of $[3, 5)$

Write in interval notation the set $\{x : 0 \le x \le 3\}$.

The set of points larger than or equal to 3 is denoted by $[3, \infty)$. The sign "∞" stands for "infinity". The set of points strictly smaller than 4 is denoted by $(-\infty, 4)$.

What is the notation for the set of numbers strictly larger than 3? Draw it on the line.

Write in interval notation the set of points smaller or equal to 0.

Write in interval notation the set of points belonging to both $(-\infty, 4)$ and $[3, \infty)$.

De Morgan's Rules

Despite his French sounding name, August De Morgan was an English mathematician, who was born in India in 1806. He made deep contributions to mathematics, but he is mainly remembered for two very simple rules that he formulated. To exemplify the first, imagine a meteorologist who predicts that tomorrow it will rain in city A and it will hail in city B. What can you deduce from the fact that he was proved wrong? That at least one of his predictions failed. Either it did not rain in A, or it did not hail in B.

Rule 1: Not "p and q" means "not p or not q".

The other rule is a mirror image. If the meteorologist predicted that it will either rain in A, or hail in B, proving him wrong means that neither occurred:

Rule 2: Not "p or q" means "neither p nor q".

If a father promised his son ice cream and chocolate, what can you deduce from the fact that he has not fulfilled his promise?

The Complement of a Set

The "complement" of a set A of points is the set of points not in A. For example, the complement of $(-\infty, 4)$ is $[4, \infty)$, since "not to be strictly smaller than 4" means "being larger than or equal to 4".

The complement of a set A is denoted by A^C.

Find $[4, \infty)^C$ and the complement of $[4, \infty)^C$.

Is there any element in the set $(-\infty, \infty)^C$?

Write its complements of the sets $\{x : x > 5\}$ and $\{x : x \ge 5\}$, and draw the complements on the line.

The Complement of an Interval

What is the complement of $[3, 4]$? $[3, 4]$ is the set of points x satisfying $x \geq 3$ and $x \leq 4$. Its complement is the set of points not satisfying both conditions, so by De Morgan's rule it is the set of points satisfying either $x < 3$ or $x > 4$. Here it is in drawing:

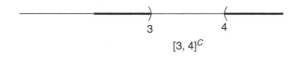

$$[3, 4]^C$$

Drawing of $[3, 4]^C$

This is also denoted by "$(-\infty, 3) \cup (4, \infty)$", where the "$\cup$" sign stands for "union", namely "taken together".

For each of the following sets write its complement, and draw the complement on the line: a. $(-2, 2]$ b. $(0, 1) \cup (3, 4)$ c. $(-\infty, 2) \cup (2, \infty)$.

Solving Inequalities

Good News (In Fact, No News)

You will be probably happy to learn that you already know how to solve inequalities. It is just like solving equations, with just one small additional rule.

Solve the inequality $80 + x > 100$.

Solution: Subtracting 20 from both sides we get $x > 20$.

From this example you have learnt the two basic facts about solving inequalities:

1. "Solving" an inequality means writing an equivalent inequality, in which the unknown is isolated, namely, it is "alone".
2. The rules for isolation are the same as in equations (well, almost...). Adding or subtracting the same number to both sides of the inequality does not change the inequality. Multiplying... well, wait a bit with that.

Joe's aunt doubled the amount of money he had, and his uncle added $400. What can you say about the amount he had if you know that now has enough money to buy a bike costing $1500?

Solution: calling the amount he had x, we know that $2x + 400 \geq 1500$. We deal with this just as with an equation:

$$2x + 400 \geq 1500\backslash - 400$$

$$2x \geq 1500 - 400 = 1100\backslash : 2$$

$$x \geq 1100 : 2 = 550$$

The last line is the solution to the inequality. The set of values satisfying the inequality is $[550, \infty)$.

Solve: a. $x + 1 \leq 2$ b. $x + 1 > 2$ c. $10x + 70 < 100$.

Multiplying an Inequality by a Negative Number

And now, to the only essential difference from equations: when you multiply the two sides of an inequality by a negative number, the inequality is reversed. Remember? The negative numbers are a "mirror world". The fact that $7 > 5$ means that $(-5) > (-7)$. Owing 5 dollars is better than owing 7.

Solve $-2x + 100 > 0$.

Solution:

$$-2x + 100 > 0\backslash - 100$$

$$-2x > -100\backslash \times -\frac{1}{2}$$

$$x < 50.$$

In the last step, the inequality was reversed. We could also avoid multiplying by a negative number, by adding $2x$ to both sides. Adding the same number to both sides does not change the inequality direction:

$$-2x + 100 > 0\backslash + 2x$$

$$100 > 2x\backslash : 2$$

$$50 > x.$$

Solve: a. $-x < -2$ b. $-x > -3$ c. $-2x > 4$.

Inequalities Involving Absolute Value

What values of x satisfy $|x - 5| < 2$? We know that $|x - 5|$ is the distance between x and 5, so these are the numbers (points) whose distance from 5 is smaller than 2. It is easy to see that these are the points between 3 and 7, namely $(3, 7)$ (open on both sides, since the inequalities are strict).

But how do you solve an inequality involving absolute values in general? The secret is in remembering that $|y| = y$ if $y \geq 0$, and $|y| = -y$ if $y \leq 0$.

Solve systematically $|x - 5| < 2$.

Solution: Divide into two cases:

Case A: $x - 5 \geq 0$. In this case $x \geq 5$ (this is what $x - 5 \geq 0$ means), and the inequality $|x - 5| < 2$ means that $x - 5 < 2$, which is the same as $x < 7$. So, in this case the set of solutions is $[5, 7)$.

Case B: $x - 5 \leq 0$. Then $x \leq 5$ (this is what $x - 5 \leq 0$ means), and since in this case $|x - 5| = -(x - 5)$ the inequality $|x - 5| < 2$ reads $-(x - 5) < 2$, or $-x + 5 < 2$, and moving sides this means $x > 3$. So, in this case the set of solutions is $(3, 5]$.

The entire set of solutions is $(3, 5] \cup [5, 7)$, which is $(3, 7)$ (drawing the sets on the line will help you realize this).

Solve: a. $|x + 4| > 10$ b. $|-2x + 10| \leq 1$.

Life becomes more complicated when two absolute sums appear in the inequality. For example:

Solve: $|x - 3| < |x + 4|$.

Thinking geometrically about this inequality makes the solution easy. $|x - 3|$ is the distance of x from 3, and $|x + 4|$ is the distance of x from (-4). So, the inequality says "the point x is closer to 3 than to (-4)". The points satisfying this are obviously the points to the right of the middle between 3 and (-4), namely to the right of $\frac{3-4}{2} = -\frac{1}{2}$. So, the solution is $x > -\frac{1}{2}$, or in set notation $\left(-\frac{1}{2}, \infty\right)$.

How to solve it systematically? Again, by dividing to cases. This time we have to divide into 4 cases:

Case A: $x - 3 \geq 0$ and $x + 4 \geq 0$. In fact this means just $x \geq 3$ (the second inequality follows from the first). In this case the inequality is $x - 3 < x + 4$. But this is always true. So, this case is just $x \geq 3$.

Case B: $x - 3 \geq 0$ and $x + 4 \leq 0$. This is impossible — why?

Case C: $x - 3 \leq 0$ and $x + 4 \geq 0$. The first inequality is $x \leq 3$, and the second is $x \geq -4$, so we have $-4 \leq x \leq 3$. In this case the inequality $|x - 3| < |x + 4|$ is: $-(x - 3) < x + 4$, or $-x + 3 < x + 4$, which after moving sides becomes $2x > -1$, meaning $x > -\frac{1}{2}$. That gives the answer $-\frac{1}{2} < x \leq 3$ in this case.

Case D: $x - 3 \leq 0$ and $x + 4 \leq 0$. This means $x \leq -4$. The inequality $|x - 3| < |x + 4|$ is then $-(x - 3) < -(x + 4)$. This is also impossible — why?

So, putting case A and C together, we have $x > -\frac{1}{2}$, as we found before.

Solve: a. $|2x| \leq |x|$ b. $|2x| < |x + 1|$ c. $|2x + 1| < 3(x + 1)$.

Inequalities That are Always True

In this section we shall meet a few inequalities that are true for every value of the variables. For example, $a + 1 > a$ is always true.

Is the inequality $2a \geq a$ true for every value of a?

(Answer — no. It is true only for non-negative a.)
Also, $x^2 \geq 0$ is true for every value of x. A square is always non-negative.

Prove that $a^2 - 4a + 4 \geq 0$ for every value of a.

Solution: $a^2 - 4a + 4 = (a - 2)^2$, and a square is always non-negative.

The Averages Inequality

Putting $x = a - b$ in the inequality $x^2 \geq 0$ gives $(a - b)^2 \geq 0$. Opening parentheses, this is $a^2 - 2ab + b^2 \geq 0$, which after moving sides reads

$$(*) \quad a^2 + b^2 \geq 2ab.$$

Next divide by 2:

$$(**) \quad \frac{a^2 + b^2}{2} \geq ab.$$

This is a nice inequality by itself, but let us do one more step. Call a^2 by a new name, say u, and give b^2 the name v. Then u and v are non-negative, and $a = \sqrt{u}$, $b = \sqrt{v}$. Putting the new names in (*) gives:

$$(***) \quad \frac{u + v}{2} \geq \sqrt{u}\sqrt{v} = \sqrt{uv}.$$

This is a famous inequality, called the "averages inequality". The reason for this name is that it says that the arithmetic average, which is the usual average, namely the middle between the two numbers, is at least as large as the geometric average. The "geometric average" of two non-negative numbers u and v is the root of their product, $\sqrt{uv}$. It is called "average" because it lies somewhere in between u and v. In particular, when $u = v$ the geometric average is the common value of u and v.

Part 15

Geometric Sequences

A man is offered two possibilities of pay for his work. In the first, he gets $10,000 every week. In the second, he gets one cent in the first week, 2 cents in the second week, 4 cents on the third, 8 cents on the fourth, and so on. Which option should he choose, assuming that he is planning on working for one year?

Those who remember the story of the Persian Shah and the inventor of the game of Chess, should know the answer. The second choice will make the man much richer. The reason is that the power operation is very strong. On the 52-nd week the man will get 2^{51} cents (not 2^{52}, just as in the first week he gets 2^0, namely 1 cent, and not 2^1 cents), and 2^{51} cents is more than 100 million dollars.

The sequence of payments (in cents) in the second option is: $1\,2\,4\,8\,16\,32\ldots$ The terms are doubled in each step. A sequence in which each term is equal to its predecessor times a fixed number is called a "geometric sequence". This fixed number is called the "quotient" of the sequence, and is often denoted by q. For example, the above sequence has quotient 2.

Here are some more examples:

1 10 100 1000 10000 (the quotient is 10)
10 30 90 270 810 (the quotient is 3)
$1 \ \frac{1}{2} \ \frac{1}{4} \ \frac{1}{8} \ \frac{1}{16}$ (the quotient is $\frac{1}{2}$).

The terms are sometimes divided by commas.

In each of the following sequences determine whether the sequence is geometric or not, and if it is find its quotient:
a. $1, 1, 1, 1, 1$ b. $2, 1, 2, 1, 2, 1$ c. $0, 0.1, 0.01, 0.001, 0.0001$.

Geometric Sequences and Powers

At the base of a geometric sequence lie powers. For example, the sequence 1, 2, 4, 8, 16, 32 is nothing but the powers of 2, starting at power 0 and ending at power 5. The sequence 10, 30, 90, 270, 810 is in fact: 10×3^0, 10×3^1, 10×3^2, 10×3^3, 10×3^4.

In general, call the first element of the sequence a_1. The second element is then the first times q, which is $a_1 q$. The third is $a_1 q^2$, and so forth. The n-th term is:

$$a_n = a_1 q^{n-1}.$$

Write the first 5 terms in a geometric sequence, in which the first element is 10, and its quotient is $\frac{1}{2}$. What is its 10-th element?

Infinite Geometric Sequences and Fractals

When a sequence is infinite, we denote it by three dots, like: $1\,2\,4\,8\,16\,32\ldots$, or $1\,\frac{1}{2}\,\frac{1}{4}\,\frac{1}{8}\,\frac{1}{16}\cdots$

Infinite geometric sequences have a very uncommon property: they are similar to their part. Take for example the sequence $1\,\frac{1}{2}\,\frac{1}{4}\,\frac{1}{8}\,\frac{1}{16}\ldots$, and multiply every element by $\frac{1}{2}$. Multiplying a system by a constant results in a similar system. But look at what we get: $\frac{1}{2}\,\frac{1}{4}\,\frac{1}{8}\,\frac{1}{16}\,\frac{1}{32}\ldots$, which is the part of the sequence starting from the second element. So, the sequence is similar to this part. In fact, multiplying it by $\frac{1}{4}$ shows that it is similar also to the sequence, $\frac{1}{4}\,\frac{1}{8}\,\frac{1}{16}\,\frac{1}{32}\ldots$, which is the part of it starting at the third place.

A system that is similar to a part of itself is called in mathematics a "fractal". You may have heard the name, and now you know an example. An infinite geometric sequence is the simplest example of a fractal. An infinite sequence of Babushkas, embedded sequentially, is another, though of course not to be found in reality. Another phenomenon in reality that if continued infinitely would be a true fractal is the shape of coasts. From afar you see a certain pattern in the tongues of land going into the sea. If you get nearer, you see on each of these tongues a similar pattern of smaller tongues.

A drawing of a fractal picture

The Sum of a Geometric Sequence

The Sum of an Infinite Geometric Sequence

A turtle is at distance 1 meter from his sweetheart. He starts walking towards her, and in the first hour he advances $\frac{1}{2}$ meter. Being tired, in the second hour he traverses only $\frac{1}{4}$ meter, in the next hour $\frac{1}{8}$ meter, and so on. Will he ever reach his beloved?

The answer is "no". But he will get as near as he wishes, because every hour he halves the distance. After one hour he is $\frac{1}{2}$ a meter away from his goal, after two hours $\frac{1}{4}$ away, and so forth. We can say that if he walks "for ever" he will reach his goal, and mathematically we write it this way:

$$\frac{1}{2} + \frac{1}{4} + \frac{1}{8} + \frac{1}{16} + \cdots = 1$$

Of course, there is no real meaning to "going for ever". It means just that if we sum more and more terms, the sums will approach 1.

Actually, we have already encountered another example of a sum of an infinite geometric sequence. We know that $0.99999\ldots$ is equal to 1. The meaning is that the sequence $0.9, 0.99, 0.999, 0.9999\ldots$ approaches 1, which is true since the distances to 1 go to zero. They are $0.1, 0.01, 0.001, \ldots$.

Dividing the equality $0.999\cdots = 1$ by 9, we get: $0.111\cdots = \frac{1}{9}$. This means that $0.1 + 0.01 + 0.001 + \cdots = \frac{1}{9}$, namely $\frac{1}{10} + \frac{1}{100} + \frac{1}{1000} + \cdots = \frac{1}{9}$. Recalling the equality $\frac{1}{2} + \frac{1}{4} + \frac{1}{8} + \frac{1}{16} + \cdots = 1 = \frac{1}{1}$ we may try to generalize. If when the quotient is $\frac{1}{2}$ the sum is $\frac{1}{1}$, and when the quotient is $\frac{1}{10}$ the sum is $\frac{1}{9}$, we may venture a guess that when the quotient is $\frac{1}{3}$ the sum is $\frac{1}{2}$, and in general if the quotient is $\frac{1}{n}$ then the sum is $\frac{1}{n-1}$, for example $\frac{1}{3} + \frac{1}{9} + \frac{1}{27} + \frac{1}{81} + \cdots = \frac{1}{2}$.

This is indeed true, and the way to see it is through the "fractal" property of geometric sequences. Take a geometric sequence $1, q, q^2, q^3 \ldots$ and sum its elements. Let S be the infinite sum $1 + q + q^2 + q^3 + q^4 + \cdots$. Multiply the sequence by q. Of course, the sum is also multiplied by q, so

$$qS = q(1 + q + q^2 + q^3 + \cdots) = q + q^2 + q^3 + q^4 + \cdots.$$

But the last sum is almost S: it is $S - 1$, since it is the sum as in S, with the first element 1 missing. So:

$$qS = S - 1,$$

which yields upon moving sides $1 = S(1-q)$, that means:

$$S = \frac{1}{1-q}.$$

Call T the sum $q + q^2 + q^3 + \cdots$. As already noted, $T = qS$, so:

$$q + q^2 + q^3 + q^4 + \cdots = \frac{q}{1-q}.$$

Substituting $q = \frac{1}{10}$ gives $T = \frac{1/10}{9/10} = \frac{1}{9}$, as we already know. Putting $q = \frac{1}{2}$ gives $T = \frac{1/2}{1/2} = 1$, which we also know.

Use the formula to show that $\frac{1}{3} + \frac{1}{9} + \frac{1}{27} + \frac{1}{81} + \cdots = \frac{1}{2}$.

Remark: all this is meaningful under one condition — that $|q| < 1$. Otherwise, the sum is not a finite number. For example, for $q = 1$ we get $1 + 1 + 1 + \cdots$, which is infinity.

The Sum of a Finite Geometric Sequence

Remember the story of the inventor of Chess? We calculated there and found that:

$$1 + 2 + 4 + 8 + \cdots + 2^{63} = 2^{64} - 1.$$

This is an example of calculation of the sum of a finite geometric sequence. Let us now calculate such sums in general, namely a sum of the form $S = 1 + q + q^2 + q^3 + \cdots + q^n$.

We shall use the same trick as in the infinite case, which is to multiply both sides by q. We get $qS = q + q^2 + q^3 + \cdots + q^n + q^{n+1}$. But look: the expression on the right hand side is almost S, the difference being that the "1" disappeared, and was replaced by the last term, q^{n+1}. So, $qS = S + q^{n+1} - 1$. Moving sides gives $S(q-1) = q^{n+1} - 1$, which transforms into:

$$S = \frac{q^{n+1} - 1}{q - 1}.$$

Recalling what S is, this means $1 + q + q^2 + q^3 + \cdots + q^n = \frac{q^{n+1}-1}{q-1}$. Putting $q = 2$ and $n = 63$ gives the result $1 + 2 + 4 + 8 + \cdots + 2^{63} = 2^{64} - 1$.